Practical
Math

GRADE 4

Kwang S. Ko, Ph.D.

1. *A SELF-STUDY GUIDE*
2. *EXERCISES*
3. *SELF-TESTS*
4. *SOLVING PROBLEMS*
5. *FULL ANSWER KEY*

Request for information should be addressed to 7722 Camino Noguera, San Diego, CA 92122
Visit our website at www.iqmaths.com

ISBN: 978-1523363025

Printed in the United States of America

10 9 8 7 6 5 4 3 2

CONTENTS

CHAPTER 1
Number Sense

In this chapter, you will learn number sense in order to compute addition, subtraction, multiplication, and division expressed in the whole numbers.

1. Place Value

1–1. A) Place Value

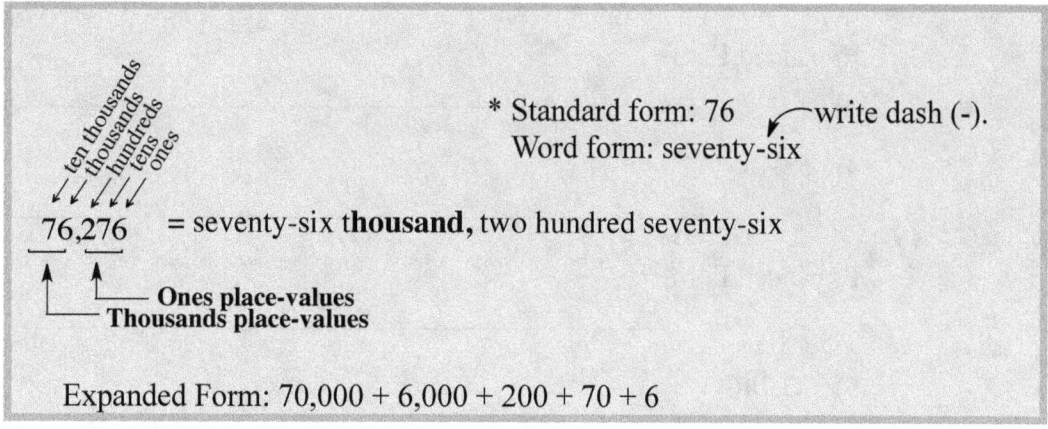

Expanded Form: $70,000 + 6,000 + 200 + 70 + 6$

B) A comma between each period can make large numbers easier to understand.

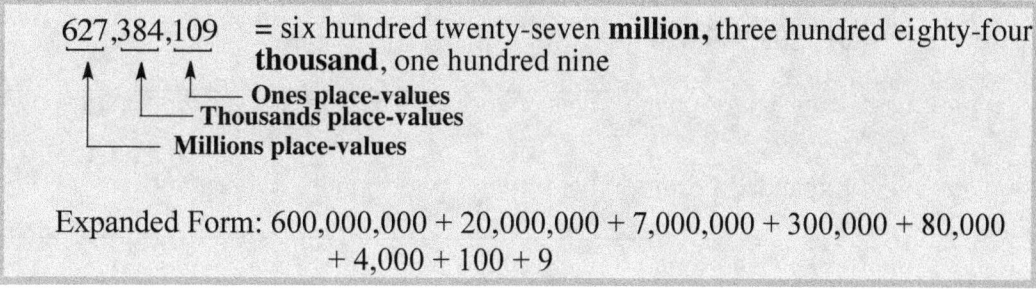

Expanded Form: $600,000,000 + 20,000,000 + 7,000,000 + 300,000 + 80,000 + 4,000 + 100 + 9$

1–2. The three forms of numbers.

a) Standard Form: 32,091,284
b) Word Form: Thirty-two million, ninety-one thousand, two hundred eighty-four
c) Expanded Form: $30,000,000 + 2,000,000 + 90,000 + 1,000 + 200 + 80 + 4$

Exercises 1   Write each number in its word and expanded forms.

1) 37

_____

Expanded Form

_____

2) 906

_____

Expanded Form

_____

3) 4,003

_____

Expanded Form

_____

4) 5,048

_____

Expanded Form

_____

5) 83,746

_____

Expanded Form

_____

6) 203,999,620

_____

Expanded Form

_____

7) 520,752,463

_____

Expanded Form

_____

Exercises 2   Write each word form in its standard form.

1) fifty-three _____

2) two hundred seven _____

3) thirty hundred thousand, eight _____

4) five hundred thirty-five million, sixteen thousand, fifteen _____

5) ninety-six thousand, forty-nine _____

6) five million, four hundred thousand, two hundred three _____

7) three hundred thousand, eight hundred eleven _____

8) two hundred seventy-seven thousand, five hundred twenty-five _____

9) eighty-one million, one thousand, eight hundred twelve_____

10) one hundred thousand, three hundred seventy-eight _____

2. Comparing Whole Numbers

1–3.   What are the relative positions of A and C? Compare the numbers.

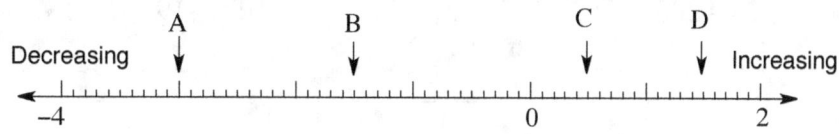

SOLUTION

The value is increasing if the number is towards the right on the number line.
The value is decreasing if the number is towards the left on the number line.
A and C are –3 and 0.5 on the line respectively. So C is greater than A.

Exercises 3   Use the number line to find the unknown values.

1) A = _____   2) B = _____

3) C = _____   4) D = _____

Exercises 4   Use the number line to find the unknown values.

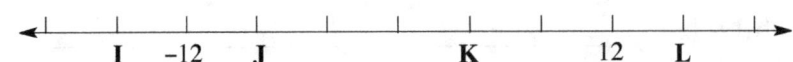

1) I = _____

2) J = _____

3) K = _____

4) L = _____

Exercises 5   Compare each pair. Use the symbols of < (greater than), = (equal to), > (less than).

1) 26 : −29

2) 0 : −3

3) −2°C : 2°C

4) 32°F : 71°F

5) 10 : −85

6) 25°C : 0°C

Exercises 6   List from greatest to least.

1) −30, −25, 0, −45, −20

2) −1°F, −7°F, −2°F, 76°F, 32°F

_____

_____

3) −200, 125, −125, −5, 200

4) −19, 25, 18, −42, −2

_____

_____

Exercises 7   List from greatest to least.

1)   532,408   534,908   503,490   89,569   195,069

_____

2)   20,746,007   27,074,007   20,900,007   3,276,007

_____

3)   892,367   829,367   982,367   893,267   892,637

_____

1–4. Comparing whole numbers

SOLUTION

a. First, compare the biggest digit with each other and determine which one is greater.
b. Second, if they are still the same number, then compare the next digit with each other and go on until there are two different numbers in the same place value.
c. Once you found the place value with different numbers, compare them to find the greatest or least digit.

2 = 2	compare to next digit ⟹	1 = 1	compare to next digit ⟹	5 > 3
21,574 ; 21,385		21,574 ; 21,385		21,574 ; 21,385

Now, you can decide which number is greater or least. So, 21,574 is greater than 21,385 and 21,385 is less than 21,574.

Exercises 8   Compare the whole numbers. Use the symbols of < (greater than), = (equal to), > (less than)

1)   1,537 ☐ 1,546   2)   96,234 ☐ 97,141

3)   67,214 ☐ 77,241   4)   7,638 ☐ 7,369

5)   28,243,648 ☐ 28,243,649   6)   1,009 ☐ 909

7)   20,203,048 ☐ 20,193,449   8)   58,914 ☐ 63,941

9)   30,201,209 ☐ 4,301,209   10)   48,736,780 ☐ 39,947,073

11)   10,200,009 ☐ 12,000,009   12)   78,336,580 ☐ 78,337,573

13)   92,214 ☐ 107,241   14)   10,201 ☐ 10,101

15)   45,027 ☐ 45,207   16)   16,234 ☐ 1,714

3. Rounding Numbers
1–5. Round Whole Numbers

> a) First, look at the place value that you want to round to.
> b) Now, you should look at the **next number** of the place value you want to round. If the place value is **5 or more**, then the rounding number **increases by 1**. If it is **less than 5**, then the rounding number **stays the same.**
>
> | * Those numbers on a line between 750 to 849 rounds up to 800 if you round to the nearest hundred. | * Those numbers on a line between 850 to 949 rounds up to 900 if you round to the nearest hundred. | * Those numbers on a line between 950 to 1,049 rounds up to1,000 if you round to the nearest hundred. |
> | --- | --- | --- |
>
> 800   900   1,000

1–6. Identify the relative position of each letter on the number line and round them to the nearest hundred.

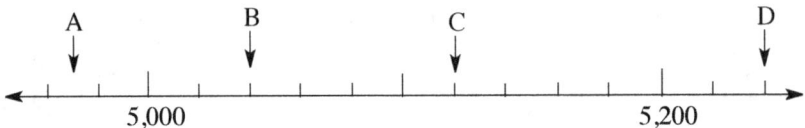

A   B   C   D

5,000   5,200

> **SOLUTION**
>
> **A** is between 4,960 and 4,980. **A** is closer to 5,000 than it is to 4,900. So it rounds to 5,000.
> **B** is 5,040. You should look at the next rounding place, which is 4. 4 is less than 5. So 5,040 rounded to the nearest hundred is 5,000.
> **C** is 5,120. 2 is less than 5. So 5,120 rounded to the nearest hundred is 5,100.
> **D** is indicated 5,240. 4 is less than 5. So 5,240 rounded to the nearest hundred is 5,200.

1–7. Round 1,673,903 to the nearest hundred thousand.

> **SOLUTION**
>
> 1,673,903
> 1) Round to the hundred thousands place, which is **6**.
> 2) Look at the next digit **7**, which is 5 or more. Therefore, **6** increases by 1.
> → 1,700,000
> └─The place value that is being rounded.

1–8. Round 5,5<u>6</u>3,985 to the nearest ten thousand.

> **SOLUTION**
>
> 5,5<u>6</u>3,985
> └─The place value that is being rounded.
>
> 1) Round to the ten thousands place, which is **6**.
> 2) Look at the next digit **3**, which is less than 5.
> Therefore, the thousands place stays the same.   ⟶   5,560,000

Exercises 9   Round each number.

1) Round 364 to the nearest ten.

_____

2) Round 8,087 to the nearest hundred.

_____

3) Round 1,054 to the nearest ten.

_____

4) Round 705,738 to the nearest ten thousand.

_____

5) Round 83,515 to the nearest thousand.

_____

6) Round 16,082 to the nearest hundred.

_____

7) Round 653,037 to the nearest ten hundred.

_____

8) Round 264,923 to the nearest ten thousand.

_____

9) Round 3,846,365 to the nearest hundred thousand.

_____

10) Round 26,148 to the nearest hundred.

_____

11) Round 72,537,205 to the nearest hundred thousand.

_____

12) Round 55,438,043 to the nearest ten million.

_____

13) Round 5,463,648 to the nearest hundred thousand.

_____

14) Round 735,267 to the nearest ten thousand.

_____

15) Round 367,736 to the nearest hundred thousand.

_____

16) Round 78,472,023 to the nearest ten million.

_____

Exercises 10   Identify the relative position of each letter on the number line.

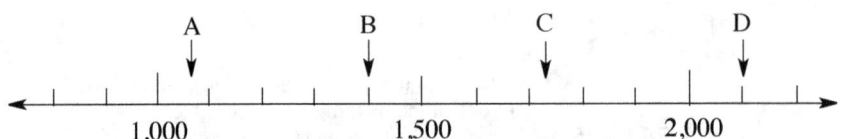

1) What is **B** rounded to the nearest thousand?

2) What number is **C** closest to?

3) Which letters are closer to 2,000? Then, which place values are rounded?

Exercises 11   Identify the relative position of each letter on the number line.

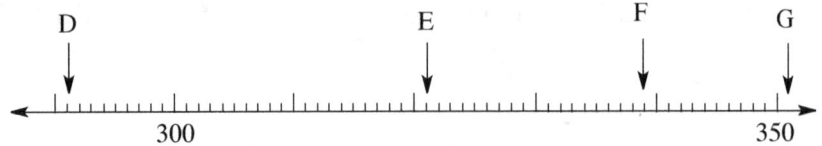

1) If **E** is rounded to 300, what is the place value that is rounded?

2) Which letter can be rounded to 400 if the value is rounded to the nearest hundred ?

3) What place value is being rounded in order to round F to 0?

Exercises 12   Identify the relative position of each letter on the number line.

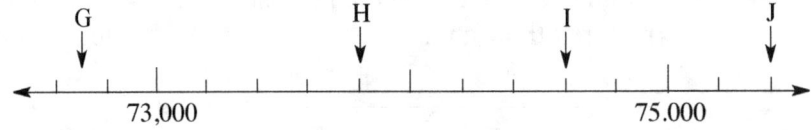

1) Name the number shown by letter **H**.

2) Name the number shown by letter **J**.

3) Which letters are closer to 70,000? Then determine the place values that are being rounded.

Exercises 13   Round each number from the place value in **bold**.

1)   1,4**9**5

2)   5,**5**63

3)   972,676

4)   24,446

5)   **6**,520

6)   6,**9**63

7)   953,909

8)   44,926

9)   143,483,394

10)   150,583,595

Exercises 14   Round to the nearest hundred.

1)   370,852

2)   6,760,952

3)   47,648

4)   46,429

5)   746,672

6)   6,509

Exercises 15   Round to the nearest hundred thousand.

1)   1,845,805

2)   5,560,879

3)   46,546,844

4)   24,837,607

5)   658,006

6)   626,450

7)   263,705

8)   37,749,059

SELF-TEST

1. Which of the following best describes the number below in word form?
    47,030

    A. forty-seven hundred, thirty
    B. forty-seven thousand, thirty
    C. forty-seven thousand, three
    D. four thousand, seven hundred, thirty

2. Which of the following best describes the number below in standard form?
    nine hundred three thousand, five hundred

    A. 9,003,500   B. 93,500
    C. 903,500   D. 905,050

3. Which of the following best describes the number below in word form?
    701,001,701

    A. seven hundred one million, one thousand, seven hundred one
    B. seventy-one million, one thousand, seven hundred one
    C. seven hundred one million, one thousand, seventy-one
    D. seven hundred one million, one hundred thousand, seven hundred one

4. Which of the following is true?

    A. 25,107 < 25,097   B. 5,000 > 40,000
    C. 10,101 is less than 10,110. D. 1,000,000 is greater than one million.

5. Which of the following is false?

    A. 402,734 ≠ four hundred two thousand, seven hundred thirty four
    B. 1,103,002 is greater than one million.
    C. 310,201 = 300,000 +10,000 + 200 + 1
    D. twenty thousand, two hundred three = 20,203

6. Which of the following is true?

    A. 104,703 < 140,097   B. 83,099 > 84,000
    C. 21,901 is less than 22,110. D. 1,000,000 is less than one million.

7. Which of the following best describes the number below in word form?
22,030,040

    A. twenty-two million, thirty thousand, forty hundred
    B. twenty-two million, thirty hundred, forty
    C. twenty-two million, thirty thousand, forty
    D. twenty-two thousand, thirty hundred, forty

8. Which of the following is false?

    A. $7,506,037 \neq$ seven million, five hundred six thousand, thirty-seven
    B. 903,002 is less than one million.
    C. $109,820 = 100,000 + 9,000 + 800 + 20$
    D. ninety-seven thousand, eight hundred four $= 97,804$

9. Which of the following is best described in standard form?
eighty thousand, six hundred

    A. 8,600   **B.** 86,000
    C. 800,600   **D.** 80,600

10. Which of the following is best described in expanded form?
60,391

    A. $60,000 + 300 + 90 + 1$ **B.** $60,000 + 3,000 + 90 + 1$
    C. $60,000 + 390 + 1$ **D.** $60,000 + 300 + 91$

11. Which of the following is best described in standard form?
$600,000 + 50,000 + 900 + 40 + 9$

    A. 659.049   **B.** 659,490
    C. 650,949   **D.** 605,949

12. What is 12,646 rounded to the nearest thousand?

    A. 12,000   **B.** 10,000
    C. 12,600   **D.** 13,000

13. What is 754,735 rounded to the nearest ten thousand?

    A. 1,000,000   **B.** 700,000
    C. 800,000   **D.** 750,000

14. What place value was rounded so that the number is 701,000?

   A. rounded to the nearest hundred thousand
   B. rounded to the nearest ten thousand
   C. rounded to the nearest thousand
   D. rounded to the nearest ten

15. What place value was rounded so that the number is now 1,000,000?

   A. rounded to the nearest million
   B. rounded to the nearest hundred thousand
   C. rounded to the nearest ten thousand
   D. possibly all of them

16. The tallest mountain in Africa, Mount Kilimanjaro is 19,341 ft high. What is the height of the mountain rounded to the nearest ten thousand?

   A. 19,000 ft   **B.** 10,000 ft
   C. 20,000 ft   **D.** 19,300 ft

17. 1 mile is equal to 5280 ft. Which of the following is 5280 rounded to the nearest thousand?

   A. 5,000 ft   **B.** 5,500 ft
   C. 5,300 ft   **D.** 6,000 ft

18. What is 96,482 rounded to the nearest ten thousand?

   A. 96,500   **B.** 90,000
   C. 96,000   **D.** 100,000

19. What is 1,909 rounded to the nearest thousand?

   A. 1,900   **B.** 2,000
   C. 1,000   **D.** 2,900

* Use the number line for Exercises **20-22**.

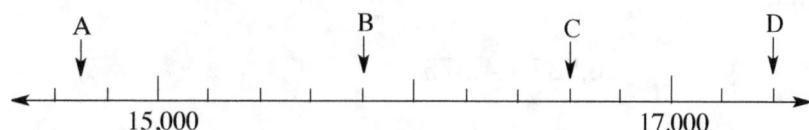

20. What is B's relative position?

    A. 15,400   **B.** 15,800
    C. 15,900   **D.** 16,000

21. Which of the following letters is the relative position of 16,600?

    A. A   **B.** B
    C. C   **D.** D

22. Which of the following letter(s) is (are) 20,000 rounded to the nearest thousand?

    A. A, B, C, and D   **B.** B, C, and D
    C. C and D   **D.** D

* Use the number line for Exercises **23-25**.

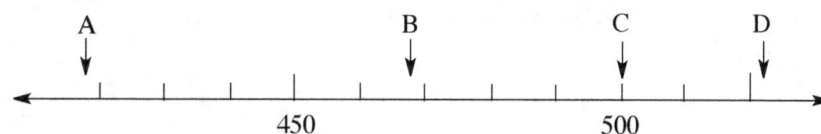

23. What is B's relative position?

    A. 488   **B.** 480
    C. 478   **D.** 468

24. Which place value of D is rounded so that the number is 1,000?

    A. Rounded to the nearest one.   **B.** Rounded to the nearest ten.
    C. Rounded to the nearest hundred.   **D.** Rounded to the nearest thousand.

25. Which of the following letter(s) is (are) 400 rounded to the nearest hundred?

    A. A   **B.** A and B
    C. A, B, and C   **D.** A, B, C, and D

4.   Estimating Sums and Differences

1–9.   Estimate.

$$6{,}537 + 4{,}375$$

SOLUTION

| $\begin{aligned}6{,}537 \\ +\ 4{,}375\end{aligned}$ | i) The thousands place value is the greatest digit.
 ii) Round to the nearest thousand.
 iii) Add. $\longrightarrow$ | $\begin{aligned}7{,}000 \\ +\ 4{,}000 \\ \hline 11{,}000\end{aligned}$ |

So, the estimated sum of 6,537 + 4,375 is about 11,000.

1–10.   Estimate.

$$64{,}528 - 16{,}485$$

SOLUTION

| $\begin{aligned}64{,}528 \\ -\ 16{,}485\end{aligned}$ | i) The ten thousands place value is the greatest digit.
 ii) Round to the nearest ten thousand.
 iii) Subtract. $\longrightarrow$ | $\begin{aligned}60{,}000 \\ -\ 20{,}000 \\ \hline 40{,}000\end{aligned}$ |

So, the estimated difference of 64,528 – 16,485 is about 40,000.

Exercises 16   Estimate each sum or difference.

1)   4,082 + 6,349   2)   5,403 – 4,821

_____   _____

3)   42,837 + 54,439   4)   36,538 – 20,947

_____   _____

5)   62,837 + 74,439   6)   46,538 – 40,307

_____   _____

7)   1,583 + 7,449   8)   2,483 – 1,821

_____   _____

9)   83,647 +104,657   10) 72,393 – 18,801

_____   _____

Exercises 17 Estimate each sum or difference to the nearest thousand.

1) $73,637 + 24,239$

2) $53,538 - 40,401$

3) $32,463 + 6,268$

4) $7,459 - 3,703$

5) $52,082 + 46,349$

6) $125,403 - 74,821$

7) $4,493 + 63,600 + 34,357$

8) $46,538 - 40,307$

9)
$$\begin{array}{r} 4,092 \\ - 1,902 \\ \hline \end{array}$$

10)
$$\begin{array}{r} 5,571 \\ + 8,430 \\ \hline \end{array}$$

11)
$$\begin{array}{r} 9,693 \\ + 8,302 \\ \hline \end{array}$$

Exercises 18 Estimate each sum or difference to the nearest ten thousand.

1) $652,137 + 423,764$

2) $362,653 - 288,038$

3) $104,682 + 83,267$

4) $159,367 - 95,730$

5) $65,081 + 44,349$

6) $704,955 - 98,676$

7)
$$\begin{array}{r} 44,692 \\ - 16,035 \\ \hline \end{array}$$

8)
$$\begin{array}{r} 23,924 \\ - 7,403 \\ \hline \end{array}$$

9)
$$\begin{array}{r} 62,728 \\ + 27,236 \\ \hline \end{array}$$

10)
$$\begin{array}{r} 76,692 \\ - 32,635 \\ \hline \end{array}$$

11)
$$\begin{array}{r} 74,308 \\ 66,736 \\ + 32,640 \\ \hline \end{array}$$

12)
$$\begin{array}{r} 5,387 \\ 14,823 \\ + 17,485 \\ \hline \end{array}$$

5. Adding and Subtracting Whole Numbers

1-11. Find the sum.

$$6,726 + 2,907$$

SOLUTION

When adding numbers, they must be lined up starting at the ones place value.

$$\begin{array}{r} 6,726 \\ +\ 2,907 \\ \hline \end{array}$$

i) Begin to add the numbers that are lined up with each other starting with the ones place value.

$$\begin{array}{r} \overset{1}{} \\ 6,726 \\ +\ 2,907 \\ \hline 3 \end{array}$$

ii) Add the ones (6 + 7) and carry the 1 to the tens column.

$$\begin{array}{r} \overset{1}{} \\ 6,726 \\ +\ 2,907 \\ \hline 3 \end{array}$$

iii) Add the tens (1 + 2).
iv) Add the hundreds (7 + 9) and carry the 1 to the thousands column.
v) Add the thousands (1 + 6 + 2).

$$\begin{array}{r} \overset{1}{}\ \overset{1}{} \\ 6,726 \\ +\ 2,907 \\ \hline 9,633 \end{array}$$

So, the sum of 6,726 and 2,907 is 9,633.

1-12. A) When adding numbers, they must be lined up starting at the ones place value.

1) Not Correct

$$\begin{array}{r} 6,726 \\ +\ 2,907 \\ \hline \end{array}$$

← line up the numbers at the ones digit.

2) Not Correct

$$\begin{array}{r} 6,726 \\ +\ 2,907 \\ \hline \end{array}$$

← line up the numbers at the ones digit.

3) Correct

$$\begin{array}{r} 6,726 \\ +\ 2,907 \\ \hline \end{array}$$

B) Even if the numbers are switched, the sum will be the same.

1) Correct

$$\begin{array}{r} 6,726 \\ +\ 2,907 \\ \hline 9,633 \end{array} = \begin{array}{r} 2,907 \\ +\ 6,726 \\ \hline 9,633 \end{array}$$

2) Correct

$$6,726 + 2,907 = 2,907 + 6,726$$

Exercises 19   Add and round to the nearest whole number.

1) $130 + 276$

2) $1,302 + 2,776$

_____

3) $218 + 260 + 233$

4) $730 + 361$

_____

5) $284 + 724 + 631$

6) $254 + 763 + 191$

_____

7) $3,218 + 260 + 7,333$

8) $751 + 2,401 + 3,432$

_____

9) $9,025 + 978$

10) $1,730 + 7,361$

_____

11) $254 + 2,763 + 1,091$

12) $6,284 + 2,684 + 9,631$

_____

13)
$$\begin{array}{r} 4.438 \\ + 2,075 \\ \hline \end{array}$$

14)
$$\begin{array}{r} 4,582 \\ + \ \ 1,509 \\ \hline \end{array}$$

15)
$$\begin{array}{r} 3,882 \\ + 2,604 \\ \hline \end{array}$$

16)
$$\begin{array}{r} 2,563 \\ + 2,415 \\ \hline \end{array}$$

17)
$$\begin{array}{r} 1,817 \\ + \ \ 4,505 \\ \hline \end{array}$$

18)
$$\begin{array}{r} 3,641 \\ + 15,255 \\ \hline \end{array}$$

1–13. Find the sum.

$$272,647 + 65,574$$

SOLUTION

$$
\begin{array}{r}
272,647 \\
+\ 65,574 \\
\hline
\end{array}
$$

i) Begin to add the numbers that are lined up with each other starting with the ones place value.

$$
\begin{array}{r}
\overset{1}{2}72,647 \\
+\ 65,574 \\
\hline
1
\end{array}
$$

ii) Add the ones (7 + 4) and carry the 1 to the tens column.

iii) Add the tens (1 + 4 + 7) and carry the 1 to the hundreds column.
iv) Add the hundreds (1 + 6 + 5) and carry the 1 to the thousands column.
v) Add the thousands place value (1 + 2 + 5).

$$
\begin{array}{r}
\overset{1}{2}72,647 \\
+\ 65,574 \\
\hline
1
\end{array}
$$

$$
\begin{array}{r}
\overset{1}{2}\overset{1}{7}2,\overset{11}{6}47 \\
+\ 65,574 \\
\hline
338,221
\end{array}
$$

vi) Add the ten thousands (7 + 6) and carry the 1 to the hundred thousands column.
vii) Add the hundred thousands place values (1 + 2).

So, the sum of 272,647 and 65,674 is 338,221.

Exercises 20   Find the sum.

1) $265,054 + 41,763 + 71,091$ 2) $31,218 + 260 + 27,333$

_____   _____

3) $543,751 + 204,601 + 24,432$ 4) $423,752 + 72,835 + 512,246$

_____   _____

5) $3,751 + 2,409 + 7,432$ 6) $2,752 + 72,835 + 2,246$

_____   _____

7)   8)   9)
$$
\begin{array}{r}
1,036 \\
+\ 3,805 \\
\hline
\end{array}
$$
$$
\begin{array}{r}
91,837 \\
+\ \ 3,848 \\
\hline
\end{array}
$$
$$
\begin{array}{r}
63,641 \\
+\ 35,376 \\
\hline
\end{array}
$$

10)   11)   12)
$$
\begin{array}{r}
83,039 \\
74,186 \\
+\ 8,685 \\
\hline
\end{array}
$$
$$
\begin{array}{r}
4,326 \\
785 \\
+\ 5,837 \\
\hline
\end{array}
$$
$$
\begin{array}{r}
45,403 \\
34,033 \\
+\ 17,029 \\
\hline
\end{array}
$$

Exercises 21   Find the sum.

1)   3,485,074 + 626,358

2)   370,474 + 849,548

3)   208,477 + 2,088,633

4)   35,026 + 75,086

5)   954,564 + 232,748

6)   844,578 + 377,687

7)
$$\begin{array}{r} 20,835 \\ + \; 2,277 \\ \hline \end{array}$$

8)
$$\begin{array}{r} 65,047 \\ + \; 55,395 \\ \hline \end{array}$$

9)
$$\begin{array}{r} 175,373 \\ + \; 335,845 \\ \hline \end{array}$$

Exercises 22   Find each value of Δ.

1) $81 + \Delta = 103$

2) $36 + \Delta = 40$

3) $\Delta + 63 = 304$

4) $\Delta + 347 = 700$

5) $150 + \Delta = 203$

6) $\Delta + 2{,}561 = 7{,}903$

7) $182 + \Delta = 459 - 185$

8) $\Delta + 451 = 963 - 234$

9) $6{,}457 + \Delta = 9{,}606 - 1{,}350$

10) $\Delta + 2{,}773 = 9{,}651 - 2{,}045$

1–14. Find the difference.

$$7{,}528 - 4{,}731$$

SOLUTION

7,528 − 4,731	i) First, the two numbers must be lined up with each other starting with the ones place value.
	ii) Subtract the ones (8 − 1).

7,528
− 4,731
           7

7,528
− 4,731
        7
iii) Subtract the tens (2 − 3), but not enough to subtract the tens. So regroup from the 1 hundred to the 2 tens = 12 tens and write 4 to the hundred.

4 12
7,5̷2̷8
− 4,731
        97
iv) Subtract the tens (12 − 3).

4 12
7,5̷2̷8
− 4,731
       97
v) Subtract the hundreds (4−7) but because there is a smaller number on top, regroup by borrowing 1000 from the thousands column so that now the hundreds column reads as (14 − 7) while the thousands place value of the top number reads as "6".

vi) Subtract the thousands (6 − 4).

6 14 12
7,5̷2̷8
− 4,731
   2,797

So, the difference of 7,528 and 4,731 is 2,797.

1–15. When you are setting up two numbers, they must be lined up at the ones digits.

1) Not correct	2) Not correct	3) Correct
7,528 − 4,731 ← line up the numbers at the ones digit.	7,528 − 4,731 ← line up the numbers at the ones digit.	7,528 − 4,731

1–16. Unlike adding, if the two numbers are switched, the difference will not be the same.

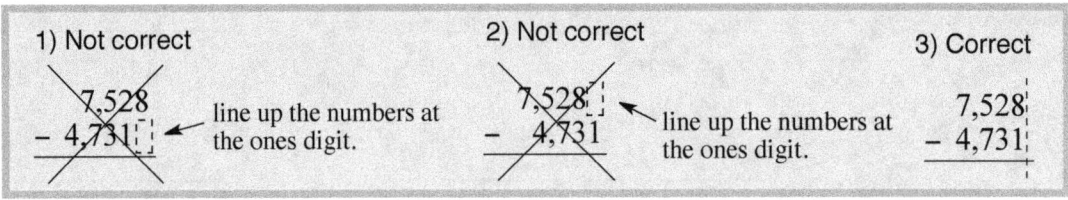

1) Not the same.	2) Not the same.
7,528 − 4,731 ≠ 4,731 − 7,528 2,697 −2,797	$7{,}528 - 4{,}731 \neq 4{,}731 - 7{,}528$

Exercises 23   Find the difference.

1)   742 – 349   2)   928 – 619

_____   _____

3)   3,052 – 2,163   4)   1,284 – 1,028

_____   _____

5)   3,218 – 769   6)   1,620 – 871

_____   _____

7)   8)   9)
$$\begin{array}{r} 768 \\ -\ 233 \\ \hline \end{array}$$ $$\begin{array}{r} 482 \\ -\ 385 \\ \hline \end{array}$$ $$\begin{array}{r} 7,529 \\ -\ 635 \\ \hline \end{array}$$

Exercises 24   Find the difference.

1)   2,751 – 2,422   2)   2,072 – 825

_____   _____

3)   5,208 – 1,369   4)   2,020 – 1,071

_____   _____

5)   3,051 – 1,462   6)   6,284 – 928

_____   _____

7)   8)   9)
$$\begin{array}{r} 440 \\ -\ 150 \\ \hline \end{array}$$ $$\begin{array}{r} 1,528 \\ -\ 849 \\ \hline \end{array}$$ $$\begin{array}{r} 2,406 \\ -\ 665 \\ \hline \end{array}$$

1–17. Find the difference.

$$627,528 - 84,731$$

SOLUTION

627,528 – 84,731	i) Subtract the ones (8 – 1).

$\overset{4\ 12}{627,\cancel{5}28}$
$-\ \ 84,731$
$\overline{\qquad 97}$

ii) Subtract the tens (2–3) but because there is a smaller number on top, regroup by borrowing 100 from the hundreds column so that now the tens column reads as (12–3), while the hundreds place value on the top number is now "4".

$\overset{4\ 12}{627,\cancel{5}28}$
$-\ \ 84,731$
$\overline{\qquad 97}$

iii) Subtract the hundreds (4–7) but because there is a smaller number on top, regroup by borrowing 1000 from the thousands column so that now the hundreds column reads as (14 – 7) while the thousands place value of the top number reads as "6".

iv) Subtract the hundred (13 – 7).
v) Subtract the thousands (6 – 4).

$\overset{14}{\overset{6\ \cancel{4}12}{627,\cancel{5}28}}$
$-\ \ 84,731$
$\overline{\quad 2,797}$

vi) Subtract the ten thousands (2 – 8) but because there is a smaller number on top, regroup by borrowing 100000 from the hundred thousands column so that now the ten thousands column reads as (12 – 8) while the hundreds thousands place value of the top number reads as "5".

vii) Subtract the hundred thousands (5 – 0).

$\overset{14}{\overset{6\ \cancel{4}12}{627,\cancel{5}28}}$
$-\ \ 84,731$
$\overline{\qquad 697}$

$\overset{14}{\overset{5\ 126\ \cancel{4}12}{\cancel{627},\cancel{5}28}}$
$-\ \ \ 84,731$
$\overline{\quad 542,797}$

So, the difference of 627,528 and 84,731 is 542,797.

Exercises 25   Find the difference.

1)   9,742 – 8,349   2)   1,528 – 629

_____   _____

3)   1,304 – 376   4)   107,052 – 57,163

_____   _____

5)   6)   7)
    21,031   53,548   52,812
  – 2,807   – 25,479   – 2,445

Exercises 26   Find the difference.

1) 31,218 − 28,269   2) 545,620 − 26,771

3) 7,052 − 163   4) 36,272 − 7,285

5) 290,702 − 8,349   6) 110,528 − 61529

7) (306,272 − 18,725) − 157,755   8) 321,751 − (205,402 − 55,714)

9)   10)   11)
    15,281   66,442   415,002
  − 6,613   − 34,479   − 4,942

Exercises 27   Find each value of Δ.

1) $76 - \Delta = 34$ 2) $63 - \Delta = 48$

3) $\Delta - 37 = 56$ 4) $\Delta - 9{,}050 = 3{,}871$

5) $43 - \Delta = 24$ 6) $\Delta - 31 = 69$

7) $5{,}306 - \Delta = 1{,}002 + 2{,}037$ 8) $\Delta - 2{,}603 = 4{,}236 - 1{,}163$

* Solving Problems

Exercises 28   Solve each problem using the given information.

1) There are 52,736 people in a baseball stadium during the Saturday game and 56,023 people during the Sunday game.
a) Estimate how many people there were at the stadium for the weekend.
b) Estimate how many more people were at the Sunday game than the Saturday game.

2) At a construction site, 17,130 bricks are ordered. But only 8,453 bricks are used. Estimate the number of bricks not used at the construction site.

Exercises 29   Solve each problem using the given information.

1) Elizabeth needs 3,745 pieces of peppermint candy with white and red colors for decoration. She found that there are four times as many white candy than red candy in a package. How many pieces of white peppermint candy does she have? Show your work.

2) Calvin and Tyler are playing a video game. Calvin's score was 1325 and Tyler's score is Calvin's score doubled. What is the sum of their score? Show your work.

3) Rick picked 738 apples. His family picked 5 times as many apples than Rick. What is the total of apples picked?

4) Corey had 644 eggs. He accidently breaks half the eggs. If he uses 246 eggs to bake the cakes, how many eggs does he have remaining?

5) At a recycling center, they recycled 45,356 bottles over a period of two days. If they recycled 27,438 bottles on the first day, how many bottles did they recycle on the second day?

6) Daniel spent $53 to buy clothes at the mall. If Daniel originally had $107, how much money does he have now? Round to the nearest whole number.

SELF-TEST

1. What is the best estimate of the sum below?
    $$125,438 + 74,803$$

    A. 200,000   B. 200,241
    C. 190,000   D. 170,000

2. What is the best estimate of the difference below?
    $$54,637 - 36,452$$

    A. 10,000   B. 15,000
    C. 19,000   D. 20,000

3. What is the best estimate of the difference below?
    $$5,293 - 2,612$$

    A. 3,000   B. 2,700
    C. 2,000   D. 2,600

4. The bakery is baking loaves of banana bread. They use 17,725 bananas for the weekdays and 14,036 bananas for the weekend. Estimate the total sum of bananas used

    A. 20,000   B. 31,000
    C. 30,000   D. 40,000

5. Which expression shows the best estimate below?
    $$33,647 - 4,450$$

    A. $33,000 - 4,000$ B. $30,000 - 4,000$
    C. $30,000 - 4,500$ D. $34,000 - 4,000$

6. Which of the following is the best estimate of $2,659 + 9,026$?

    A. 11,000   B. 12,000
    C. 10,000   D. 11,600

7. Which of the following expressions represents the sum of 7,037 and 763?

    A. $7,037 \div 763$ B. $7,037 \times 763$
    C. $7,037 - 763$ D. $7,037 + 763$

8. Which of the following expressions represents the difference of 7,037 and 763?

    A. $7,037 - 763$ **B.** $7,037 \div 763$

    C. $7,037 \times 763$ **D.** $7,037 + 763$

9. What is the value of the following expression?
$$\$938 + \$5,363$$

    A. 70,000   **B.** 60,000

    C. 54,301   **D.** 64,301

10. What is the value of the following expression?
$$\$239.00 + \$809.00$$

    A. 11,084,139   **B.** 2,490,048

    C. 2,500,000   **D.** 2,490,000

11. What is the value of the following expression?
$$14,025 - 5,026$$

    A. 58,999   **B.** 68,999

    C. 58,765   **D.** 59,000

12. A store is having a sale on bowties. They had 53,014 bowties in storage. By the end of the month, they had sold 14,378 bowties. How many bowties do they have left?

    A. 420   **B.** 50

    C. 64   **D.** 399

13. A bakery store used 1,178 eggs in the morning to bake their products. The bakers used 368 eggs to make chocolate chip cookies and used 505 eggs to bake cakes. How many eggs do they have left?

    A. 55   **B.** 31

    C. 516   **D.** 3.6

14. The science club needs $4,142 in order to buy new equipment so they are asking for donations. The club decides to sell tickets for a fundraiser. They sold 364 tickets for $873. People donated twice as much as the club sold tickets. How much more money is needed before the science club can reach their goal?

A. $1,507,688   **B.** $4,142
C. $11   **D.** $3,269

15. Jason finished reading 437 pages of a book so far. He read half of the pages in the book. How many pages are in the book?

   A. 650   **B.** 874
   C. 655.5   **D.** 218.5

16. What is the best value of the following expression?
$$15,370 - 7,628$$

   A. 7,762   **B.** 7,748
   C. 7,752   **D.** 7,742

17. What is the best value that fits in the box?
$$3,407 + \boxed{} = 11,029$$

   A. −7,622   **B.** 14,436
   C. 7,622   **D.** 3

18. What is the value of Δ for the expression below?
$$\Delta - 8,027 = 1,074$$

   A. 6,953   **B.** 9,101
   C. −6.953   **D.** 8

19. What is the value of Δ for the expression below?
$$1,321 - \Delta = 326$$

   A. 995   **B.** 1,647
   C. −995   **D.** 4

6. Estimating Products and Quotients

1–18. Estimating multiplication and division problems

$$
\begin{array}{r} 548 \\ \times\ \ 7 \\ \hline \end{array}
$$

i) The hundred-place value is the greatest digit.
ii) Round to the nearest hundred.
iii) Multiply. ⟶

$$
\begin{array}{r} 500 \\ \times\ \ 7 \\ \hline 3500 \end{array}
$$

So the estimated product of 548 × 7 is about 3500.

$$
\begin{array}{r} 48 \\ \times\ 47 \\ \hline \end{array}
$$

i) The ten-place value is the greatest digit.
ii) Round to the nearest ten for both numbers.
iii) Multiply. ⟶

$$
\begin{array}{r} 50 \\ \times\ 50 \\ \hline 2500 \end{array}
$$

So the estimated product of 48 × 47 is about 2500.

1–19. Estimating the quotient is similar to estimating the product.

$9\overline{)643}$

i) The hundred-place value is the greatest digit,
ii) Divide the hundreds (6 ÷ 9), but not enough. So, divide the tens (64 ÷ 9). The dividend of 64 can divide between 63 and 72 by 9. ⟶
iii) Write zero for the rest of digit.

$9\overline{)643}^{\,70}$ or $9\overline{)643}^{\,80}$

9 x 7 = 63
9 x 8 = 72

So, the estimated quotient 643 ÷ 9 is about 70 and 80.

- - - Numbers and Names - - - - - - - -

24 ÷ 8 = 3

↑ dividend   ↑ divisor   ↖ quotient

divisor ⟶ $8\overline{)24}^{\,3}$ ← quotient

↑ dividend

Exercises 30   Estimate each quotient and product.

1)   49 × 2   2)   83 ÷ 7

_____   _____

3)   69 × 5   4)   67 × 8

_____   _____

5)   25 ÷ 7   6)   54 ÷ 6

_____   _____

Exercises 31   Estimate each quotient and product.

1) 3×48 **2)** $23 \div 7$

_____   _____

3) 259×5 **4)** 4×247

_____   _____

5) $925 \div 4$ **6)** $2254 \div 6$

_____   _____

7) 302×7 **8)** 6×370

_____   _____

9) $1,024 \div 4$ **10)** $5,040 \div 6$

_____   _____

Exercises 32   Estimate each quotient and product.

1)
$$558 \times 3$$

2)
$$9 \times 5409$$

3)
$$2\overline{)824}$$

4)
$$4 \times 7075$$

5)
$$653 \times 7$$

6)
$$6\overline{)424}$$

7)
$$5\overline{)420}$$

8)
$$6\overline{)628}$$

9)
$$8\overline{)4940}$$

7.   Multiplying and Dividing Whole Numbers
1–20. Find the product.

$$78 \times 7$$

SOLUTION

When you multiply a 2-digit number by a 1-digit number, follow the steps.

$\begin{array}{r} 78 \\ \times\ 7 \\ \hline \end{array}$	i) The numbers of two groups must be line up to the ones. ii) Multiply the ones digits and carry the 5 to the tens column. $8 \times 7 = 56$	$\begin{array}{r} 5 \\ 78 \\ \times\ 7 \\ \hline 6 \end{array}$

$\begin{array}{r} 5 \\ 78 \\ \times\ 7 \\ \hline 6 \end{array}$ iii) Multipy the tens digits and then add the 5 to the product of the tens digits. $7 \times 7 = 49$ $\begin{array}{r} 5 \\ 78 \\ \times\ 7 \\ \hline 546 \end{array}$

So, the product of 78 ×7 is 546.

1–21. When multiplying, if the numbers are switched, then the product will still be the same.

1) Correct	2) Correct
$\begin{array}{r} 78 \\ \times\ 7 \\ \hline 546 \end{array} = \begin{array}{r} 7 \\ \times\ 78 \\ \hline 546 \end{array}$	$78 \times 8 = 8 \times 78$

Exercises 33   Find the product.

1) 63×3 2) 47×5

_____   _____

3) 35×5 4) 42×8

_____   _____

5) 4×92 6) 36×8

_____   _____

Exercises 34   Find the product.

1) 7×71 2) 29×2

_____   _____

3) 52×3 4) 24×7

_____   _____

5) 7×379 6) 9×442

_____   _____

7)   8)   9)
$$\begin{array}{r} 337 \\ \times\ \ 9 \\ \hline \end{array}$$ $$\begin{array}{r} 304 \\ \times\ \ 5 \\ \hline \end{array}$$ $$\begin{array}{r} 543 \\ \times\ \ 4 \\ \hline \end{array}$$

10)   11)   12)
$$\begin{array}{r} 260 \\ \times\ \ 7 \\ \hline \end{array}$$ $$\begin{array}{r} 231 \\ \times\ \ 8 \\ \hline \end{array}$$ $$\begin{array}{r} 326 \\ \times\ \ 6 \\ \hline \end{array}$$

Exercises 35   Find each value of Δ.

1) $6 \times \Delta = 78$ 2) $5 \times \Delta = 210$

_____   _____

3) $\Delta \times 7 = 532$ 4) $\Delta \times 9 = 1,368$

_____   _____

5) $15 \times \Delta = 315$ 6) $\Delta \times 3 = 1,854$

_____   _____

7) $55 \times \Delta = 880$ 8) $\Delta \times 24 = 792$

_____   _____

1–22.   Find the product.

$$269 \times 4$$

> **SOLUTION**
>
> When you multiply a 3-digit number by a 1-digit number, follow the steps.
>
> $\begin{array}{r} 269 \\ \times\ 4 \\ \hline \end{array}$ i) The ones place value of the two factors must be lined up $\begin{array}{r} ^3 \\ 269 \\ \times\ 4 \\ \hline 6 \end{array}$
>
> ii) Multiply the 9 and 4 to get 36. Since 36 is greater than 9, carry the 3 to the tens colum. $4 \times 9 = 36$
>
> $\begin{array}{r} ^3 \\ 269 \\ \times\ 4 \\ \hline 6 \end{array}$ iii) Multiply 6 and 4 and add the 3 to the product to get 27. Carry the 2 to the hundreds column. $4 \times 6 = 24$ $\begin{array}{r} ^{23} \\ 269 \\ \times\ 4 \\ \hline 1076 \end{array}$
>
> iv) Multiply the 2 and 4 and add the 2 to get 10. Since there are no more digits to carry over, simply write 10 in the front.
>
> $4 \times 2 = 8$
>
> So, the product of 269×4 is 1,076.

Exercises 36   Find the product.

1) 462×8 **2)** 305×5

_____   _____

3) $2 \times 4{,}233$ **4)** $2{,}751 \times 7$

_____   _____

5) 463×8 **6)** 904×2

_____   _____

7) 6×182 **8)** 5×196

_____   _____

9) $1{,}884 \times 8$ **10)** 6×582

_____   _____

1–23. Find the product.

$$624 \times 28$$

SOLUTION

When you multiply a 3-digit number by a 2-digit number, follow the steps.

$$\begin{array}{r} 624 \\ \times\ 28 \\ \hline \end{array}$$

i) The ones place value of the two factors must be lined up

ii) First multiply by the ones place value of the 2nd number.

$$\begin{array}{r} {\scriptstyle 1\ 3} \\ 624 \\ \times\ 28 \\ \hline 4992 \end{array}$$

$$\begin{array}{r} {\scriptstyle 1\ 3} \\ 624 \\ \times\ \ 8 \\ \hline 4992 \end{array}$$

iii) Then multiply by the tens place value of the 2nd number. Put 0 in the ones place value of the quotient and continue to multiply.

iv) Add the product together.   (624 x 20)

$$\begin{array}{r} {\scriptstyle 1\ 3} \\ 624 \\ \times\ 28 \\ \hline 4992 \end{array}$$

$$\begin{array}{r} {\scriptstyle 1\ 3} \\ 624 \\ \times\ 28 \\ \hline 4992 \\ +\ 12480 \\ \hline 17,472 \end{array}$$

$$\begin{array}{r} 624 \\ \times\ 20 \\ \hline 12480 \end{array}$$

* Put 0 in the ones place value.

A) Standard Form: 28
B) Word Form: twenty-eight
C) Expanded Form: 20 + 8

So, the product of 624 × 28 is 17,472.

Exercises 37   Find the product.

1) 36×13 **2)** 47×25

_____   _____

3) 65×54 **4)** 42×87

_____   _____

5) 559×92 **6)** 362×84

_____   _____

7)
$$\begin{array}{r} 283 \\ \times\ 51 \\ \hline \end{array}$$

8)
$$\begin{array}{r} 19 \\ \times\ 626 \\ \hline \end{array}$$

9)
$$\begin{array}{r} 245 \\ \times\ 37 \\ \hline \end{array}$$

Exercises 38   Find the product.

1) 15×201

2) 19×257

3) 734×35

4) 24×762

5) 711×39

6) 125×42

7)
$$\begin{array}{r} 507 \\ \times\ 18 \\ \hline \end{array}$$

8)
$$\begin{array}{r} 55 \\ \times\ 903 \\ \hline \end{array}$$

9)
$$\begin{array}{r} 748 \\ \times\ 29 \\ \hline \end{array}$$

Exercises 39   Find each value of Δ.

1) $7 \times \Delta = 63$

2) $4 \times \Delta = 96$

3) $\Delta \times 3 = 258$

4) $\Delta \times 5 = 685$

5) $5 \times \Delta = 435$

6) $\Delta \times 19 = 703$

7) $25 \times \Delta = 350$

8) $\Delta \times 16 = 5,472$

9) $6 \times \Delta = 4,956$

10) $\Delta \times 24 = 1,032$

1–24. When dividing, if the numbers are switched, then the quotient will be the difference.

1) Not Correct	2) Not Correct
$4\overline{)112} \neq 112\overline{)4}$	$112 \div 4 \neq 4 \div 112$

- Remember the terms.

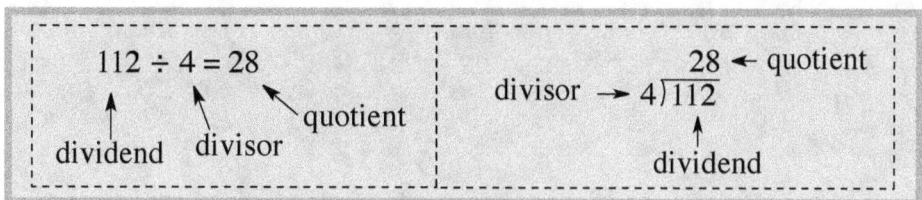

1–25. Find the quotient.

$$324 \div 6$$

SOLUTION

When you divide a 3-digit number by a 1-digit number, follow the steps.

$6\overline{)324}$

i) Determine how many times the divisor can go into the dividend and write down that number. In this case, the divisor can fit in the first two digits of the dividend.

ii) Divide the tens place value (32/6) and write 5 in the tens place value of the quotient, as 6 goes into 32 five times.
iii) Subtract 30 from 32.

$$\begin{array}{r} 5 \\ 6\overline{)324} \\ -30 \\ \hline 2 \end{array}$$ 6 x 5 = 30

$$\begin{array}{r} 5 \\ 6\overline{)324} \\ -30 \\ \hline 2 \end{array}$$

iv) Carry down the 4.
v) Divide the ones place value (24/6) and multiply the quotient by 6.
vi) Subtract 24 from 24.

$$\begin{array}{r} 54 \\ 6\overline{)324} \\ -30 \\ \hline 24 \\ -24 \\ \hline 0 \end{array}$$ 6 x 5 = 30 6 x 4 = 24

So, the quotient of 324 ÷ 6 is 54.

Exercises 40   Find the quotient.

1)   84 ÷ 3

2)   296 ÷ 8

_____

3)   632 ÷ 4

4)   165 ÷ 5

_____

Exercises 41   Find the quotient.

1) $909 \div 3$ 2) $765 \div 5$

_____   _____

3) $533 \div 4$ 4) $744 \div 4$

_____   _____

5) $684 \div 6$ 6) $315 \div 7$

_____   _____

7)   8)   9)
   $9\overline{)162}$ $19\overline{)646}$ $18\overline{)576}$

10)   11)   12)
   $28\overline{)364}$ $48\overline{)1008}$ $46\overline{)1380}$

Exercises 42   Find each value of Δ.

1) $427 \div \Delta = 7$ 2) $26 \div \Delta = 2$

_____   _____

3) $\Delta \div 7 = 52$ 4) $\Delta \div 9 = 43$

_____   _____

5) $135 \div \Delta = 15$ 6) $\Delta \div 18 = 2$

_____   _____

1–26. Find the quotient with a remainder.

$$867 \div 4$$

SOLUTION

When you divide a 3-digit number by a 1-digit number, follow the steps.

4⟌867

i) Determine how many times the divisor can go into the dividend and write down that number. In this case, the divisor can fit in the first digit of the dividend. →

$$\begin{array}{r} 2 \\ 4\overline{)867} \\ -8 \\ \hline 0 \end{array}$$ $4 \times 2 = 8$ $$\begin{array}{r} 2 \\ 4\overline{)8} \\ -8 \\ \hline 0 \end{array}$$

$$\begin{array}{r} 2 \\ 4\overline{)867} \\ -8 \\ \hline 0 \end{array}$$

ii) Divide the hundreds place value and write 2 in the hundreds place value of the quotient.
iii) Subtract 8 from 8. →
iv) Carry down the 6.
v) Divide the tens place value. As 4 only fit into 6 once, write 1 in the tens place value of the quotient.
vi) Subtract 4 from 6 below.

$$\begin{array}{r} 21 \\ 4\overline{)867} \\ -8\downarrow \\ \hline 06 \\ -4 \\ \hline 2 \end{array}$$ $4 \times 2 = 8$ $4 \times 1 = 4$ $$\begin{array}{r} 1 \\ 4\overline{)6} \\ -4 \\ \hline 2 \end{array}$$

$$\begin{array}{r} 21 \\ 4\overline{)867} \\ -8\downarrow \\ \hline 06 \\ -4 \\ \hline 2 \end{array}$$

vii) Carry down the 7.
viii) Divide the ones place value and write 6 in the ones place value of the quotient. →

ix) Subtract 24 from 27. 3 is the remainder of the equation.

$$\begin{array}{r} 216r3 \\ 4\overline{)867} \\ -8\downarrow\downarrow \\ \hline 06\downarrow \\ -4\downarrow \\ \hline 27 \\ -24 \\ \hline 3 \end{array}$$ $4 \times 2 = 8$ $4 \times 1 = 4$ $4 \times 6 = 24$ $$\begin{array}{r} 6 \\ 4\overline{)27} \\ -24 \\ \hline 3 \end{array}$$

So, the quotient of $867 \div 4$ is 216 with a remainder of 3.

- Remember the terms.

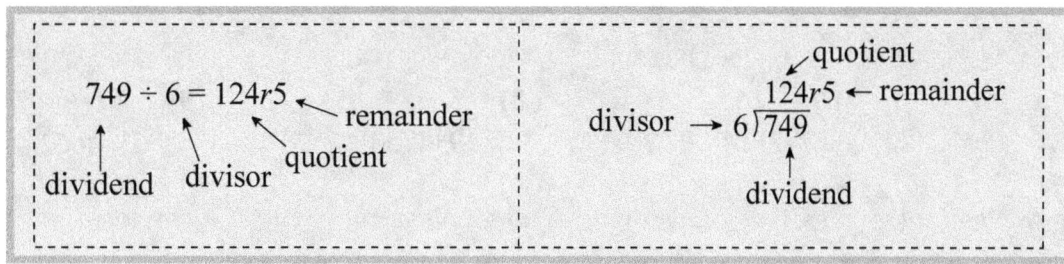

Exercises 43   Find the quotient with a remainder.

1) $256 \div 3$ 2) $736 \div 3$

_____   _____

3) $632 \div 4$ 4) $255 \div 4$

_____   _____

5) $644 \div 7$ 6) $909 \div 5$

_____   _____

Exercises 44   Find the quotient.

1) $626 \div 6$ 2) $434 \div 7$

_____   _____

3) $293 \div 5$ 4) $768 \div 6$

_____   _____

5) $486 \div 8$ 6) $1,093 \div 8$

_____   _____

7)   8)   9)
$7\overline{)2814}$ $19\overline{)4750}$ $63\overline{)4662}$

10)   11)   12)
$12\overline{)3746}$ $25\overline{)2150}$ $54\overline{)1063}$

Exercises 45   Find each value of Δ.

1) $190 \div \Delta = 8r6$

2) $32 \div \Delta = 16$

_____

_____

3) $\Delta \div 7 = 52r1$

4) $\Delta \div 9 = 68r3$

_____

_____

5) $247 \div \Delta = 35r2$

6) $\Delta \div 8 = 54r1$

_____

_____

7) $126 \div \Delta = 18$

8) $\Delta \div 4 = 92$

_____

_____

1-27.   A remainder exists when it is not enough to divide the dividend with the divisor.

The relationship of the remainder when dividing is shown in the following:

i) $749 \div 6 = 124r5$ The quotient is written with a remainder left over

ii) $749 \div 6 = 124 + \frac{5}{6}$ Write the remainder as a fraction.

iii) $(749 - 5) \div 6 = 124$ Subtract the remainder on both sides.

iv) $749 - 5 = (6 \times 124)$ Multiply 6 on both sides.

v) $744 = 744$ The remainder is true.

Exercises 46   Determine the value that goes in each box.

1)

$$12\overline{)920}$$ 76r⬚

2)

109r13
$$15\overline{)}$$

3)

25r23
$$24\overline{)}$$

4)

⬚
$$8\overline{)997}$$

5)

⬚
$$8\overline{)2782}$$

6)

26r42
$$47\overline{)}$$

7)

124r5
$$6\overline{)}$$

8)

15r7
$$10\overline{)}$$

9)

148r3
$$5\overline{)}$$

* Solving Problems

Exercises 47   Solve each problem using the given information.

1) In a school with 218, students in the 4th and 5th grades, each student is required to read 6 books per semester. Estimate how many books are read by the 4th and 5th grades per semester in total.

2) Andy has 316 marbles. If he distributes them evenly between 6 boxes, estimate how many marbles are in each box.

3) Parker wants to organize his collection of photos. He has 423 photos. If he wants to divide the photos equally between 5 albums, find how many photos are needed per album.

Exercises 48   Solve each problem using the given information.

1) A lemonade stand has 50 cups of lemonade. So far, the owners have sold 24 cups. How many cups do they have left over?

2) A plane is depositing powdered fire retardant on a forest fire. On the first run, the plane dumps powder over 42 acres. On the second run, the plane dumps over twice as many acres than the first run. How many acres did the plane dump powder over in total?

3) In 2010, Farmer Carter grew 146,984 cabbages. In 2011, he grew 94,748 cabbages more than in 2010. How many more cabbages did he grow in total?

4) A table is set out for a buffet. On the table are a few plates of shrimp. Each plate has exactly 152 pieces of shrimp. There are 456 pieces of shrimp in total. How many plates of shrimp are set out?

5) In Lucy has $1,023 in her bank account. For two weeks, she deposited $24 each week. How much money does Lucy have in her bank account?

SELF-TEST

1. What is the estimated product of the expression below?

$$398 \times 5$$

    A. 10,000   **B.** 20,000
    C. 30,000   **D.** 40,000

2. What is the estimated quotient of the expression below?

$$609 \div 8$$

    A. 60 or 70   **B.** 60 or 80
    C. 70 or 80   **D.** 80 or 90

3. What is the estimated difference of the expression below?

$$5{,}293 - 592$$

    A. 4,400   **B.** 4,600
    C. 4,500   **D.** 4,000

4. The bakery is baking loaves of banana bread. They use 18,725 bunches of banana per week. If there are 8 bananas to a bunch, how many bananas do they have? Estimate the value

    A. 1,870   **B.** 1,000
    C. 2,000   **D.** 1,800

5. Which of the following represents 9 in the equation below?

$$872 \div \mathbf{9} = 96 + \frac{8}{9}$$

    A. Quotient   **B.** Remainder
    C. Dividend   **D.** Divisor

6. Which of the following represents $\frac{3}{4}$ in the equation below?

$$479 \div 4 = 119 + \frac{3}{4}$$

    A. Quotient   **B.** Addend
    C. Dividend   **D.** Divisor

7. Which of the following represents the product of 7,037 and 763?

       A. 7,037 ÷ 763   **B.** 7,037 × 763
       C. 7,037 − 763   **D.** 7,037 + 763

8. Which of the following represents the quotient of 7,037 and 763?

       A. 7,037 ÷ 763   **B.** 7,037 × 763
       C. 7,037 − 763   **D.** 7,037 + 763

9. What is the product of the expression below?
$$4 \times 363$$

       A. 1,232   **B.** 1,432
       C. 1,442   **D.** 1,452

10. What is the quotient of the expression below?
$$504 \div 9$$

       A. 36   **B.** 45
       C. 54   **D.** 56

11. What is the product of the expression below?
$$5 \times 126$$

       A. 530   **B.** 630
       C. 540   **D.** 640

12. At Halloween, Mr. Myers gave away 7 pieces of candy to each trick-or-treater. If 67 trick-or-treaters came to his door, how many pieces of candy did he give away?

       A. 9.6   **B.** 60
       C. 469   **D.** 74

13. A bakery has two kinds of bags. The D bag can hold half as less than the Z bag. If the Z bag has 96 bagels in total, how many bagels are in the D bag?

       A. 48   **B.** 94
       C. 192   **D.** 24

14. A tray can hold 6 eggs. If there are 1,494 eggs, how many trays will be needed?

A. 1,560   B. 1,430
C. 249   D. 23

15. At an art exhibition, nails are needed in order to hang up the paintings. If a painting
     needs 9 nails to stay on the wall and there are 162 paintings, how many nails are needed
     in total?

A. 12   B. 18
C. 153   D. 558

16. What is the product in the expression below?
     $$252 \times 7$$

A. 1,764   B. 1,564
C. 1,664   D. 1,864

17. What is the product in the expression below?
     $$6 \times 573$$

A. 3,428   B. 3,438
C. 3,038   D. 3,018

18. What is the quotient in the expression below?
     $$401 \div 7$$

A. $58 + \dfrac{2}{7}$ B. $57r3$

C. $57 + \dfrac{2}{7}$ D. $58r3$

19. What is the best value of Δ for the equation below?
     $$264 \div \Delta = 33$$

A. 5   B. 6
C. 7   D. 8

20. What is the best value of the remainder (Δ)?
     $$704 \div 6 = 117r\Delta$$

A. 2   B. 4
C. 6   D. 7

21. What is the value of Δ for the equation below?
$$82 \times \Delta = 574$$

A. 5   **B.** 6
C. 7   **D.** 8

22. Which of the following expressions exemplifies the relationship with the remainder below?
$$570 \div 24 = 23r18$$

A. $(23 \times 24) \times 18$ **B.** $(23 \times 24) + 18$
C. $(23 + 18) \times 24$ **D.** $(24 + 23) \times 18$

23. Which of the following equations exemplifies the relationship with the remainder below?
$$55 \div 4 = 13r3$$

A. $55 \div 4 = 13 + 3$ **B.** $13 + \dfrac{3}{4} = 55 \div 4$
C. $55 = (13 + 4) \times 3$ **D.** $(13 + 3) \times 4 = 55$

24. Which of the following equations does not follow the relationship with the remainder below?
$$79 \div 8 = 9r7$$

A. $(79 - 7) \div 8 = 9$ **B.** $9 + \dfrac{7}{8} = 79 \div 8$
C. $79 = (9 + 8) \times 7$ **D.** $79 \div 8 = 9 + \dfrac{7}{8}$

25. Which of the following equations does not follow the relationship with the remainder below?
$$(25 - 1) \div 4 = 6$$

A. $25 \div 4 = 6r1$ **B.** $24 \div 4 = 6 + \dfrac{1}{4}$
C. $(25 \div 4) - 1 = 6$ **D.** $25 = (4 \times 6) + 1$

CHAPTER 2
Decimals and Fractions

In this chapter, you will solve decimal and fraction problems involving addition, subtraction, multiplication, and division.

1. Place Value

2–1. Number Sense: Operations with Decimals

A. Understanding place value

a) Write the whole number in the expression, which is left of the decimal point.
b) Write "**and**" to represent the decimal point.
c) Write the decimal part, which is right of the decimal point.

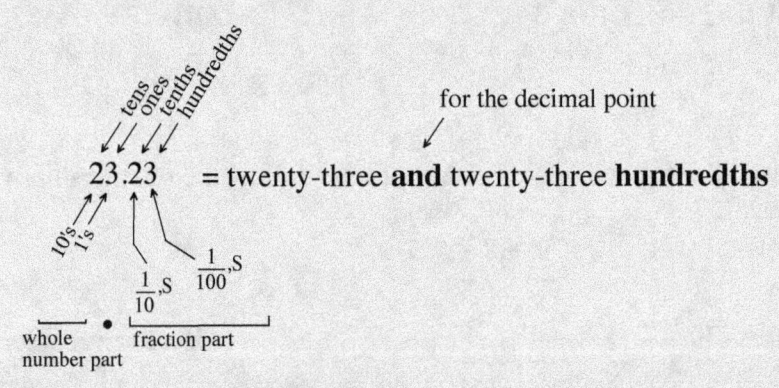

Expanded Form: $20 + 3 + \dfrac{2}{10} + \dfrac{3}{100}$

B. Names in three forms.

a) Standard Form: 384.907
b) Word Form: Three hundred eighty-four and nine hundred seven thousandths
c) Expanded Form: $300 + 80 + 4 + \dfrac{9}{10} + \dfrac{7}{1000}$

C. Understanding equivalent decimals

There are the same amount.

a) $0.8 = 0.80 = 0.800$
b) $\$3 = \$3.0 = \$3.00$

Exercises 1   Write each number in their word and expanded forms.

1)   0.01 =

_____

Expanded Form

_____

2)   0.60 =

_____

Expanded Form

_____

3)   10.08 =

_____

Expanded Form

_____

4)   19.64 =

_____

Expanded Form

_____

5)   0.50 =

_____

Expanded Form

_____

6)   4.12 =

_____

Expanded Form

_____

7)   4.012 =

_____

Expanded Form

_____

Exercises 2   Write each number of word form in its standard form.

1) nine and five hundredths _____

2) zero and sixths _____

3) twenty-six hundredths _____

4) one tenth _____

5) seven and thirty-five hundredths _____

6) sixty-one and two tenths _____

7) zero and eighty-one hundredths _____

8) one and eleven hundredths _____

2. Comparing Numbers

2–2. Comparing numbers

a. First, compare the same tenth digit of both numbers and determine if they are different.
b. Second, if they are still the same number, then compare the hundredths digits with each other. If they are identical, then continue until you find a place value where the digits are different.
c. Once you found the different numbers, compare them to determine the greatest and least of the digits.

2–3. Compare the decimals.

$$1.75 \ \boxed{} \ 1.71$$

SOLUTION

First, compare the ones digits of both numbers. If they are the same number, compare the tenth digit with each other. If they are still the same number then compare the hundredths digits.

$1 = 1$	compare to next digit $\Rightarrow$	$.7 = .7$	compare to next digit $\Rightarrow$	$0.05 > 0.01$
1.75 $\boxed{}$ 1.71		1.75 $\boxed{}$ 1.71		1.75 $\boxed{>}$ 1.71

Now, you can decide which of the decimals is greater. So, 1.75 is greater than 1.71 or 1.71 is less than 1.75.

Exercises 3   Compare the decimals with each other.

1)   8.78 _____ 8.09   **2)**   0.2 _____ 0.21

3)   14.082 _____ 14.82   **4)**   84.19 _____ 84.099

5)   0.44 _____ 0.24   **6)**   2 _____ 1.999

7)   4.082 _____ 4.820   **8)** $\frac{1}{5}$ _____ 0.210

9)   0.26 _____ $\frac{1}{4}$ **10)** $2\frac{1}{10}$ _____ 1.999

11)   2.000 _____ −1.999   **12)** $\frac{1}{10}$ _____ −0.10

Exercises 4   List from greatest to least.

1)   2.05, 2.07, 2.15, 2.17, 2.27   **2)**   0.359, 0.38, 0.392, 0.387, 0.383

_____   _____

3)   2.648, 2.6489, 2.939, 1.998   **4)**   0.3, 0.35, 0.6, 0.65, 0.05

_____   _____

5)   0.26, $\frac{1}{5}$, 0.245, $\frac{1}{4}$ **6)**   0.3, $\frac{1}{3}$, 0.303, $\frac{3}{100}$, 0.34

_____   _____

7)   10.01, 0.99, 8.90, 9.01   **8)**   2.85, 2.36, 2.65, 2.56, 2.58

_____   _____

9)   −0.03, 0.05, 0.10, −0.05   **10)** 0.11, 0.22, −0.11, 0.05, −0.22

_____   _____

Exercises 5   a) Use the number lines to identify each unknown value.

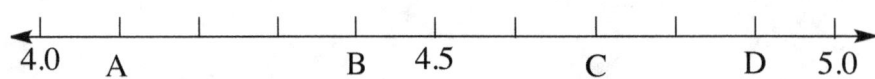

1) A = _____   **2)** B = _____

3) C = _____   **4)** D = _____

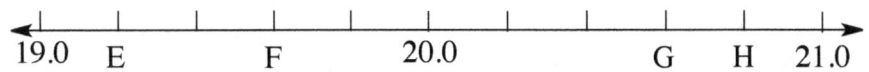

5) E = _____   **6)** F = _____

7) G = _____   **8)** H = _____

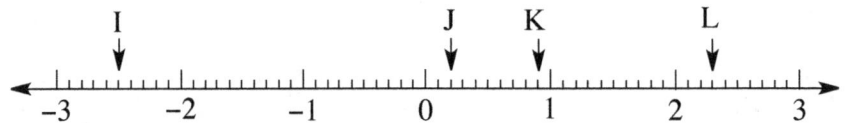

9) I = _____   **10)** J = _____

11) K = _____   **12)** L = _____

b) Mark each letter on the number line.

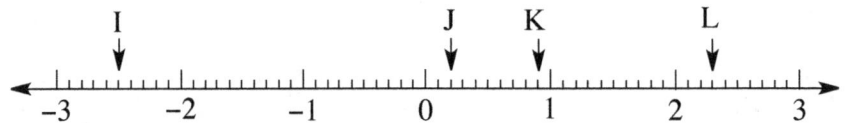

13) M = −0.05   **14)** N = 0.08

15) O = 0.25   **16)** P = 0.45

SELF-TEST

1. Which of the following is the correct word form for the decimal below?
   1.19

   A. one and nineteen tenths
   C. one, nineteen hundredths

   B. one and nineteen hundredths
   D. one, nineteen tenths

2. Which of the following is the correct expanded form for the decimal below?
   1.11

   A. $1 + \dfrac{1}{100}$

   C. $1 + \dfrac{1}{10} + \dfrac{1}{100}$

   B. $\dfrac{1}{10} + \dfrac{1}{100} + \dfrac{1}{1000}$

   D. $1 + \dfrac{11}{10} + \dfrac{1}{100}$

3. Which of the following is the correct standard form for the decimal below?
   $$3 + \dfrac{4}{10}$$

   A. 3.4
   C. 30.40

   B. 3.04
   D. 3.004

4. Which of the following is the correct standard form for the decimal below?
   four and eleven hundredths

   A. 4.10
   C. 4.11

   B. 41.0
   D. 4.01

5. Which of the following is incorrect?

   A. $1 < 1.000$
   C. 0.27 is less than 0.272.

   B. $2.370 > 2.037$
   D. 0.2 is greater than 0.02.

6. Which of the following is correct?

   A. $1.28 \neq$ one and twenty-eight hundredths
   C. $\dfrac{8}{100} = 0.008$

   B. 0.002 is greater than 0.200.
   D. 2.49 equals to $2\dfrac{49}{100}$.

* Use the number line for Exercises **7-9**.

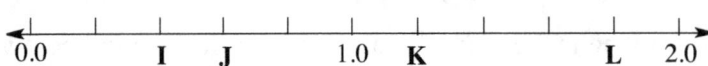

7. What is the relative position of I?

    A. 0.20   **B.** 0.30
    C. 0.40   **D.** 0.60

8. Name the relative position of 1.2.

    A. I   **B. J**
    C. K   **D. L**

9. Which of the following letters are 2.0 rounded to the nearest whole number?

    A. I   **B. J**
    C. L   **D. J, K**, and **L**

* Use the number line for Exercises **10-12**.

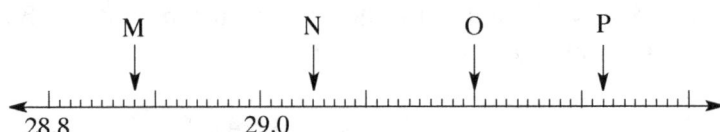

10. What is the relative position of M?

    A. 28.9   **B.** 28.88
    C. 29.12   **D.** 29.88

11. What is the relative position of O?

    A. 31.03   **B.** 30.30
    C. 29.33   **D.** 29.40

12. Name the relative position of 29.32.

    A. M   **B. N**
    C. O   **D. P**

3. Adding and Subtracting Decimals

2–4. Add the decimals.

$$2.47 + 0.9$$

SOLUTION

Line up the decimal points when you add.

$$\begin{array}{r} 2.47 \\ + 0.9 \\ \hline \end{array}$$

i) Line up the decimal places.
ii) Write zeros if necessary.

iii) When writing the answer, put the decimal point in the same place.

$$\begin{array}{r} 2.47 \\ + 0.90 \\ \hline \end{array}$$

iv) Add them as if they are whole numbers.

$$\begin{array}{r} {}^{1} \\ 2.47 \\ + 0.90 \\ \hline 3.37 \end{array}$$

added 0, but the value did not change.

So, the value of 2.47 + 0.9 is 3.37.

There are equivalent decimals.

$$0.9 = 0.90 = 0.900$$

Exercises 6   Add the decimals.

1)   1 + 0.8   **2)**   0.6 + 0.5

_____   _____

3) $1.05 + $0.25   **4)**   0.2 + 1.9

_____   _____

5)   0.9 + 0.2   **6)** $2.90 + $3.40

_____   _____

7)   2.6 + 1.4   **8)**   0.6 + 8.5

_____   _____

9)
$$\begin{array}{r} 0.41 \\ + \ 0.9 \\ \hline \end{array}$$

10)
$$\begin{array}{r} 1.6 \\ + \ 0.65 \\ \hline \end{array}$$

11)
$$\begin{array}{r} \$0.55 \\ + \ \$1.45 \\ \hline \end{array}$$

12)
$$\begin{array}{r} \$1.05 \\ + \ \$0.68 \\ \hline \end{array}$$

13)
$$\begin{array}{r} 0.69 \\ + \ 7.43 \\ \hline \end{array}$$

14)
$$\begin{array}{r} 15.0 \\ + \ \ 1.05 \\ \hline \end{array}$$

2-5.   Find the value of Δ.

$$1.6 + \Delta = 2.4$$

SOLUTION

For solving Δ, you should subtract 1.6 from both sides.

$$1.6 - \mathbf{1.6} + \Delta = 2.4 - \mathbf{1.6}$$

The left side of the numbers can be canceled out (1.6 – 1.6 = 0) and then only Δ remains. The right side of the numbers is subtracted (2.4 – 1.6 = 0.8). So Δ is 0.8.

Exercises 7   Find each value of Δ.

1) $0.8 + \Delta = 1.3$ 2) $0.02 + \Delta = 0.15$

_____   _____

3) $\Delta + 1.2 = 3.4$ 4) $\Delta + 0.37 = 0.93$

_____   _____

5) $1.50 + \Delta = 2.03$ 6) $\Delta + 2.5 = 3.8$

_____   _____

7) $\Delta + 0.06 = 1.09$ 8) $\Delta + \$5.6 = \$9.35$

_____   _____

Exercises 8   Add the decimals.

1) $0.25 + 0.75$ 2) $5.02 + 5.18$

_____   _____

3) $\$6.50 + \$2.75$ 4) $0.74 + 0.27$

_____   _____

5) $0.05 + 0.8$ 6) $\$1.05 + \$0.75$

_____   _____

Exercises 9   Add the decimals.

1) $9.06 + 1.8$

2) $0.02 + 2.2$

3) $0.3 + 0.09$

4) $\$14.90 + \$0.99$

5) $12.07 + 9.44 + 0.79$

6) $1.76 + 5.9 + 2.49$

7)
```
  7.69
  2.05
+ 0.5
```

8)
```
  0.46
  0.2
+ 0.57
```

9)
```
  $5.05
  $0.50
+ $2.77
```

Exercises 10   Find each value of Δ.

1) $3.6 + \Delta = 5.9$

2) $2.03 + \Delta = 3.21$

3) $\Delta + 0.09 = 0.41$

4) $\Delta + 5.37 = 8.93$

5) $\$4.68 + \Delta = \$8.03$

6) $\Delta + 0.25 = 0.48$

7) $2.45 + \Delta = 5.9 - 1.85$

8) $\Delta + 1.05 = 4.66 - 2.34$

9) $6.57 + \Delta = 8.6 - 1.35$

10) $\Delta + 2.77 = 5.65 - 1.04$

2-6. Subtract the decimals.

$$1.02 - 0.6$$

SOLUTION

Line up the decimal points when you subtract.

$$\begin{array}{r} 1.02 \\ -\ 0.6 \\ \hline \end{array}$$

i) Line up the decimal places.
ii) Write zeros if necessary.

iii) When writing the answer, put the decimal point in the same place.

$$\begin{array}{r} 1.02 \\ -\ 0.60 \\ \hline \end{array}$$

iv) Subtract them as if they are whole numbers.

added 0, but the value did not change.

$$\begin{array}{r} {}^{0}\ {}^{10} \\ \cancel{1}.\cancel{0}2 \\ -\ 0.60 \\ \hline 0.42 \end{array}$$

Exercises 11   Find the difference of the decimals

1) $1 – $.25

2)   3 – 0.15

3)   0.6 – 0.2

4)   0.4 – 0.1

5)   0.7 – 0.2

6)   0.8 – 0.3

7)   0.5 – 0.04

8)   0.2 – 0.03

9)   5.02 – 1.44

10) $3.15 – $1.25

11) $1.50 – $0.75

12)   0.53 – 0.24

13)   10.07 – 1.55

14)   12.84 – 3.28

15)
$$\begin{array}{r} 4.5 \\ -\ 1.05 \\ \hline \end{array}$$

16)
$$\begin{array}{r} \$1.25 \\ -\ \$0.99 \\ \hline \end{array}$$

17)
$$\begin{array}{r} 0.93 \\ -\ 0.05 \\ \hline \end{array}$$

2–7. Find the value of Δ.

$$\Delta - 0.7 = 1.3$$

SOLUTION

For solving Δ, you should add 0.7 to both sides.

$$\Delta - 0.7 + \mathbf{0.7} = 1.3 + \mathbf{0.7}$$

The left side of the numbers can be canceled out $(0.7 - 0.7 = 0)$ and then only Δ remains. The right side of the numbers can be added $(1.3 + 0.7 = 2.0)$. So Δ is 2.0.

Exercises 12   Find each value of Δ.

1) $1.4 - \Delta = 0.9$ 2) $4.2 - \Delta = 2.9$

_____   _____

3) $\Delta - 0.8 = 0.8$ 4) $\Delta - 2.5 = 0.52$

_____   _____

5) $\$4.90 - \Delta = \$2.03$ 6) $\Delta - 6.5 = 3.8$

_____   _____

Exercises 13   Find the difference of the decimals.

1) $\$0.73 - \$0.29$ 2) $2.07 - \$0.25$

_____   _____

3) $\$2.36 - \$.75$ 4) $\$1.05 - \$0.76$

_____   _____

5) $0.25 - 0.09$ 6) $1.09 - 0.17$

_____   _____

Exercises 14   Find the difference of the decimals

1) $0.27 – $0.09

2) $1.25 – $0.85

3)   3.07 – 0.99

4)   10.24 – 3.25

5)   0.72 – 0.28

6)   1.05 – 0.55

7) $11.75 – ($5.39 – $2.95)

8)   ($8.25 – $2.35) – $3.86

9)   (71.05 – 19.29) – 13.47

10)   21.35 – (12.02 – 4.88)

11)   2.72 – (1.28 – 0.89)

12)   (5.05 – 2.55) – 1.05

Exercises 15   Find each value of Δ.

1) $0.64 – Δ = 0.28$

2) $3.43 – Δ = 1.05$

3) $Δ – 3.56 = 1.81$

4) $Δ – 1.75 = 1.18$

5) $0.99 – Δ = 0.43$

6) $Δ – 3.92 = 5.86$

7) $12.30 – Δ = 7.43$

8) $Δ – 2.75 = 5.25$

Exercises 16   Add or subtract the decimals

1) $1.45 – $0.99   2) $0.90 + $0.25

_____   _____

3)   9.02 – 3.07   4) $0.25 – $0.05

_____   _____

5)   (15.24 – 8.55) + 2.08   6)   ($3.75 – $2.05) + $2.15

_____   _____

7)   1.85 + (1.35 – 0.98)   8) $1.25 + ($8.05 – $3.85)

_____   _____

9)   10)   11)
$$\begin{array}{r} 0.3 \\ -\ 0.05 \\ \hline \end{array}$$ $$\begin{array}{r} \$9.65 \\ +\ \$1.47 \\ \hline \end{array}$$ $$\begin{array}{r} 3.23 \\ -\ 2.85 \\ \hline \end{array}$$

Exercises 17   Find each value of Δ.

1) $2.03 – \Delta = 1.890$ 2) $0.7 + \Delta = 2.01$

_____   _____

3) $\Delta + 11.35 = 17.9$ 4) $\Delta – 1.8 = 3.14$

_____   _____

5) $16.9 + \Delta = 19.81$ 6) $\Delta + 0.42 = 1.5$

_____   _____

7) $25.01 – \Delta = 13.6$ 8) $4.3 – \Delta = 1.90$

_____   _____

1. What is the value of the expression below?
    1.05 – 0.49

    A. 70.00   **B.** 77.66
    C. 67.56   **D.** 87.66

2. Shelly has $458.36 on her account in the bank and deposited $71.85. How much money does she have in total now?

    A. $386.51   **B.** $530.21
    C. $530.00   **D.** $386.00

3. What is the sum of 2.9 + 5.33?

    A. 8.23   **B.** 8.13
    C. 8.33   **D.** 8.03

4. At a track meet, 5 runners are running the 100-meter race. Their total time is 100.8 seconds. 4 of the scores are 20.5 seconds, 19.2 seconds, 20.1 seconds, and 18.4 seconds. What is the time for the fifth runner?

    A. 24.6   **B.** 23.6
    C. 22.6   **D.** 21.6

5. You have $31. You want to buy your friend a present that costs $17.50. Additionally, you are feeling hungry and want to buy a ham sandwich, which costs $4.50. How much money will you have left over?

    A. $8.00   **B.** $9.00
    C. $10.00   **D.** $11.00

6. What is the sum of $2.03 + $0.25?

    A. $2.28   **B.** $1.78
    C. $2.18   **D.** $2.38

7. It rained 1.57 inches in the last week and rained 0.283 inches today. How much did it rain in total?

A. 158.283	**B.** 1.853
C. 1.287	**D.** 1.340

8. Wallington has a total of $46.35 in his wallet. If he used $18.50 to buy movie tickets and his parents gives him $35.20, how much money does he have now?

A. $81.55	**B.** $27.85
C. $63.05	**D.** $64.85

9. What is the value of Δ?

$$\$3.79 - \Delta = \$2.84$$

A. $6.63	**B.** $1.05
C. $0.85	**D.** $0.95

10. What is the value of Δ?

$$\$1.05 - \Delta = \$0.87$$

A. 0.19	**B.** 1.92
C. 0.18	**D.** 1.82

11. What is the value of 1.25 – 0.99?

A. 0.25	**B.** 0.26
C. 2.24	**D.** 2.14

12. What is the value of Δ?

$$\$6.03 + \Delta = \$19.22$$

A. $12.19	**B.** $24.25
C. $13.19	**D.** $25.25

13. What is the value of Δ?

$$\Delta + 0.99 = 2.28$$

A. −1.29	**B.** 3.27
C. 2.26	**D.** 1.29

4. Prime Factors, LCM, and LCD

2–8. Determine the prime factors.

> **a)** Factor- factors are numbers or quantity that when multiplied with another, produces a given number or expression.
>   For Example, $6 = 3 \times 2$. So 2 and 3 are factors of 6.
>
> **b)** Prime number (prime factor)- A prime number is a positive integer greater than 1 that has no positive divisors other than 1 and itself.
>   For Example, 2, 3, 5, 7, 11, 13, 17, and so on.
>
> **c)** Prime factorization- prime factorization is finding the set of products of the original integer.

2-9. Find the prime factors of 12.

> **SOLUTION**
>
> First, find any factors whose products equal 12 and then keep on factoring until only prime factors are remain.
>
>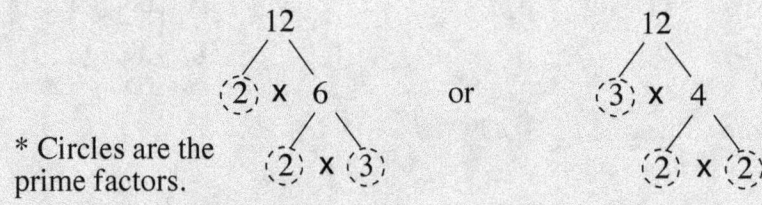
>
> * Circles are the prime factors.
>
> • The prime factorization of 12 is $2 \times 2 \times 3$ or $2^2 \times 3$.
> • The prime factors of 12 are 2, 2, and 3.

Exercises 18   Find the factors of each number.

1) 8 =   2) 34 =

3) 5 =   4) 20 =

Exercises 19   Find the prime numbers.

1) 4 =   2) 17 =

3) 71 =   4) 84 =

Exercises 20   Find the prime factorization of each number.

    1) 24 =   **2)** 18 =

    3) 27 =   **4)** 31 =

    5) 9 =   **6)** 232 =

2–10. Least Common Multiple (LCM)

When you are adding or subtracting fractions with the different denominators, you can use their least common multiple. The LCM is the smallest common multiple of both numbers. To find the common multiples of two or more number, list the multiples of each number and find the first common value.

2–11. Find the LCM of 2 and 4.

Multiples of 2: 2, **4**, 6, and so on.
Multiples of 4: **4**, 8, 12, and so on.
The LCM of 2 and 4 is 4.

2–12. Least Common Denominator (LCD)

The least common denominator is the LCM of the denominators of the fractions.

2–13. Find the LCD of $\dfrac{1}{10}$ and $\dfrac{1}{2}$.

SOLUTION

a) First, find the LCM of the denominators.

$\dfrac{1}{10}$ and $\dfrac{1}{2}$ $\begin{cases} \text{Multiples of 2: 2, 4, 6, \textbf{8}, \textbf{10}, 12, and so on.} \\ \text{Multiples of 10: \textbf{10}, 20, 30, and so on.} \end{cases}$

The LCM of the denominators, 2 and 10, is **10**.

Two different denominators

b) The LCD is the LCM of the denominators.
    So, the LCD of $\dfrac{1}{10}$ and $\dfrac{1}{2}$ is 10.

Exercises 21   Find the LCM.

    1)   12 and 7   **2)**   3 and 11

    3)   6 and 10   **4)**   21, 8, and 6

    5)   4, 9, and 18   **6)**   6, 15, and 5

Exercises 22   Find the LCD.

    1) $\frac{1}{2}$ and $\frac{2}{3}$ **2)** $\frac{4}{6}$ and $\frac{11}{12}$

    3) $\frac{2}{5}$ and $\frac{5}{6}$ **4)** $\frac{1}{2}$ and $\frac{3}{4}$

    5) $\frac{5}{7}$ and $\frac{3}{8}$ **6)** $\frac{2}{3}$ and $\frac{1}{4}$

Exercises 23   Find the LCD.

    1) $\frac{1}{4}$ and $\frac{2}{6}$ **2)** $\frac{1}{2}$ and $\frac{4}{5}$

    3) $\frac{2}{5}$ and $\frac{5}{10}$ **4)** $\frac{1}{3}$ and $\frac{10}{15}$

5. Understanding Fractions

2–14. Write a fraction that represents the shaded part of the square.

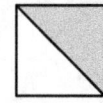

> **SOLUTION**
>
> The square is divided into two equal parts and contains one shaded part.
>
> $$\frac{\text{Number of parts in the shaded region}}{\text{Number of equal parts in all}} = \frac{1 \leftarrow \text{numerator}}{2 \leftarrow \text{denominator}}$$
>
> Read as "one half".

Exercises 24   Write a fraction that represents the shaded part of the square.

1)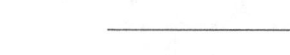

_____

2)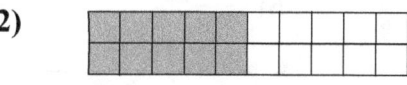

_____

3)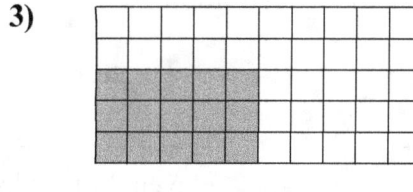

_____

4)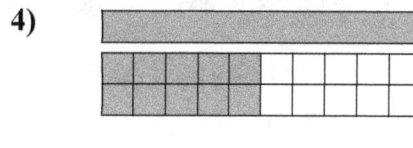

_____

5)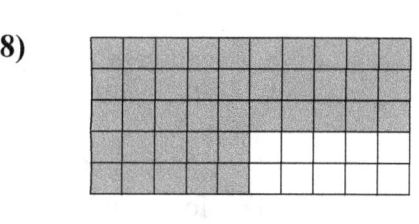

_____

6)

_____

7)

_____

8)

_____

Exercises 25   Write each fraction in its word form.

1) $\frac{1}{2}$ =

_____

2) $\frac{2}{3}$ =

_____

3) $\frac{1}{8}$ =

_____

4) $\frac{1}{4}$ =

_____

5) $\frac{1}{10}$ =

_____

6) $\frac{1}{100}$ =

_____

Exercises 26   Write each word form in its standard form.

1)   one tenth   2)   five tenths

_____   _____

3)   four sevenths   4)   eleven thirteenths

_____   _____

5)   eight ninths   6)   one hundredth

_____   _____

7)   one and one fourth   8)   two and two tenths

_____   _____

9)   three fifths   10)   nine sixths

_____   _____

11)   four hundredths   12)   seven eights

_____   _____

6. Comparing Fractions

2–15. Compare the fractions.

$$\frac{1}{3} \ \square \ \frac{2}{3}$$

SOLUTION

Given that two different fractions have the <u>same denominators</u>, then the numerators can be compared in order to compare the fractions themselves.

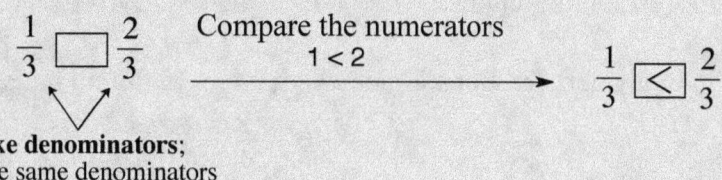

like denominators;
the same denominators

* One shaded block out of 3 blocks is smaller than two shaded blocks out of 3 blocks.

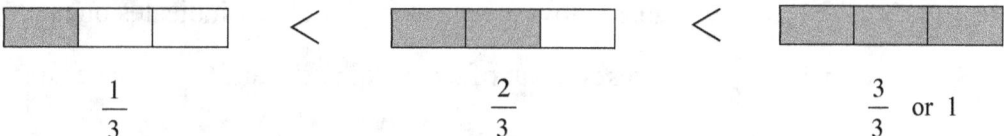

$$\frac{1}{3} \qquad\qquad \frac{2}{3} \qquad\qquad \frac{3}{3} \ \text{or} \ 1$$

2–16. If two different fractions have <u>different denominators,</u>

   a. First, find the LCD of both fractions.
   b. Second, rewrite the fractions as equivalent.
   c. Then, compare the numerators of the fractions.

2–17. Compare the fractions.

$$\frac{1}{2} \ \square \ \frac{1}{10}$$

The denominators of two fractions have 2 and 10. So you should find the LCD of the two fractions and then convert them to have like denominators.

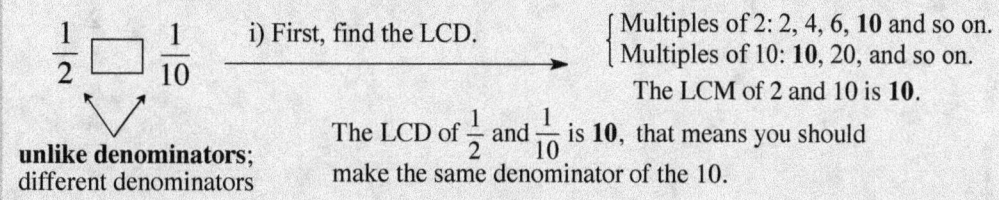

i) First, find the LCD.

Multiples of 2: 2, 4, 6, **10** and so on.
Multiples of 10: **10**, 20, and so on.

The LCM of 2 and 10 is **10**.

The LCD of $\frac{1}{2}$ and $\frac{1}{10}$ is **10**, that means you should make the same denominator of the 10.

unlike denominators;
different denominators

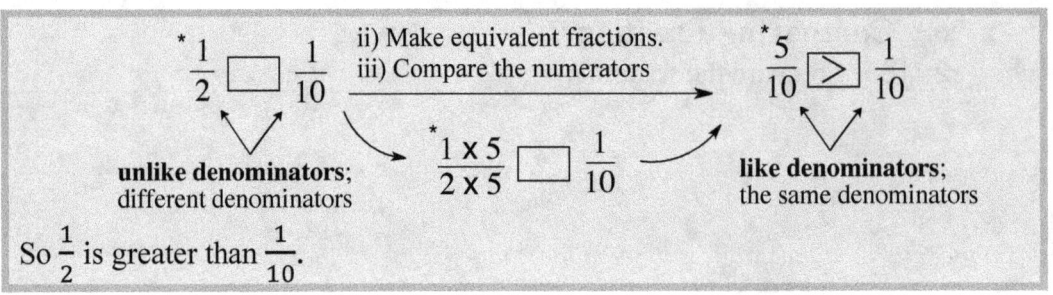

So $\frac{1}{2}$ is greater than $\frac{1}{10}$.

2–17. Equivalent fractions.

* Equivalent fractions: Fractions that look different but are equal to each other.

$$\frac{1}{2} = \frac{1 \times 5}{2 \times 5} = \frac{5}{10}$$

equivalent fractions

* From the figure below, 1 indicates a whole , $\frac{1}{2}$ indicates one out of two equal parts, $\frac{5}{10}$ indicates five out of ten equal parts, and $\frac{1}{10}$ indicates one out of ten equal parts. $\frac{5}{10}$ is equivalent to $\frac{1}{2}$.

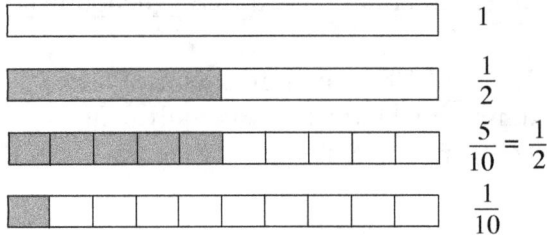

Exercises 27   Find the equivalent fraction.

1) $\frac{1}{3} = \boxed{}$

2) $\frac{1}{2} = \boxed{}$

3) $\frac{9}{12} = \boxed{}$

4) $\boxed{} = \frac{2}{5}$

5) $\boxed{} = \frac{4}{7}$

6) $\boxed{} = 1\frac{1}{2}$

Exercises 28   Find the equivalent fraction.

1) $\dfrac{3}{7} = \boxed{} = \boxed{}$

2) $1\dfrac{1}{3} = \boxed{} = \boxed{}$

3) $\dfrac{\boxed{}}{4} = \dfrac{15}{\boxed{12}} = \dfrac{\boxed{}}{36}$

4) $\dfrac{2}{\boxed{}} = \dfrac{\boxed{10}}{15} = \dfrac{30}{\boxed{}}$

5) $\dfrac{\boxed{}}{5} = 1\dfrac{3}{15}$

6) $1\dfrac{4}{6} = \dfrac{\boxed{}}{18}$

7) $\dfrac{2}{3} = \dfrac{\boxed{}}{18}$

8) $\dfrac{\boxed{}}{7} = \dfrac{20}{28}$

9) $\dfrac{5}{9} = \dfrac{\boxed{}}{18}$

10) $\dfrac{\boxed{}}{4} = \dfrac{9}{12}$

Exercises 29   Compare the fractions. Use the symbols of < (greater than), = (equal to), > (less than)

1) $\dfrac{5}{7} \;\boxed{}\; \dfrac{4}{7}$

2) $\dfrac{1}{2} \;\boxed{}\; \dfrac{1}{6}$

3) $\dfrac{6}{7} \;\boxed{}\; 1$

4) $\dfrac{3}{9} \;\boxed{}\; \dfrac{1}{3}$

5) $\dfrac{1}{2} \;\boxed{}\; \dfrac{6}{8}$

6) $1\dfrac{1}{2} \;\boxed{}\; \dfrac{12}{8}$

7) $\dfrac{2}{12} \;\boxed{}\; \dfrac{4}{6}$

8) $\dfrac{8}{10} \;\boxed{}\; \dfrac{4}{5}$

9) $1 \;\boxed{}\; \dfrac{6}{5}$

10) $\dfrac{2}{8} \;\boxed{}\; \dfrac{2}{4}$

11) $1\dfrac{3}{5} \;\boxed{}\; \dfrac{16}{10}$

12) $\dfrac{3}{4} \;\boxed{}\; \dfrac{2}{8}$

13) $\dfrac{1}{9} \;\boxed{}\; \dfrac{1}{7}$

14) $\dfrac{2}{3} \;\boxed{}\; \dfrac{8}{12}$

Name: Date:

Exercises 30   List from greatest to least.

1) $\dfrac{1}{5}, \dfrac{1}{4}, \dfrac{1}{3}, \dfrac{1}{2}$

2) $\dfrac{2}{1}, \dfrac{3}{2}, \dfrac{4}{3}, \dfrac{5}{4}$

_____

3) $\dfrac{2}{2}, \dfrac{3}{4}, \dfrac{4}{6}, \dfrac{5}{8}$

4) $\dfrac{2}{8}, \dfrac{3}{4}, \dfrac{2}{6}, \dfrac{1}{2}$

_____

5) $\dfrac{1}{8}, \dfrac{2}{3}, \dfrac{3}{4}, \dfrac{1}{2}$

6) $\dfrac{1}{3}, \dfrac{3}{5}, \dfrac{3}{4}, \dfrac{1}{2}$

_____

7) $0.25, \dfrac{2}{5}, \dfrac{1}{3}, 0.30$

8) $0.75, \dfrac{2}{3}, \dfrac{4}{7}, \dfrac{5}{8}$

_____

9) $\dfrac{4}{6}, \dfrac{3}{9}, \dfrac{2}{4}, \dfrac{5}{15}$

10) $\dfrac{4}{5}, \dfrac{3}{4}, \dfrac{2}{3}, \dfrac{1}{2}$

_____

Exercises 31   Mark each letter on the number line.

1) $A = \dfrac{2}{5}$

2) $B = \dfrac{3}{4}$

3) $C = \dfrac{4}{20}$

4) $D = \dfrac{15}{100}$

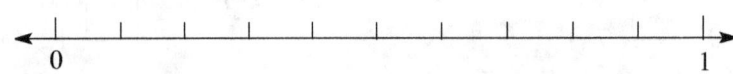

Exercises 32   Use the number line to find the unknown values.

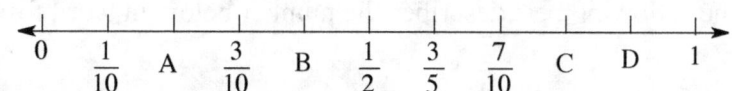

1) A = _____   **2)** B = _____

3) C = _____   **4)** D = _____

Exercises 33   Use the number line to find the unknown values.

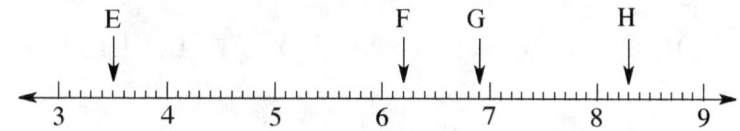

1) E = _____   **2)** F = _____

3) G = _____   **4)** H = _____

Exercises 34   Use the number line to find the unknown values.

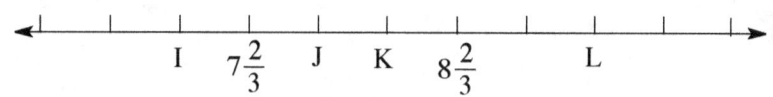

1) I = _____   **2)** J = _____

3) K = _____   **4)** L = _____

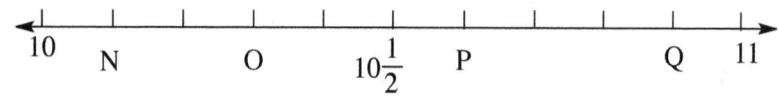

5) N = _____   **6)** O = _____

7) P = _____   **8)** Q = _____

1. Which of the following best describes the number below in word form?

$$\frac{7}{8}$$

    A. seven and eight tenths   **B.** seven, eighths
    C. seven divided by eight   **D.** seven eighths

2. Which of the following best describes the decimal below in standard form?
one hundredth

    A. $\dfrac{1}{10}$ **B.** $\dfrac{1}{100}$

    C. 1.01   **D.** 1.100

3. Which of the following best describes the decimal below in standard form?
Two and two hundredths

    A. 2.20   **B.** 220.00
    C. $\dfrac{22}{100}$ **D.** 2.02

4. Which of the following is the prime factorization for 8?

    A. $1 \times 2 \times 2 \times 4$ **B.** 2×4
    C. $2 \times 2 \times 2$ **D.** 1×8

5. What are the common factors of 8 and 14?

    A. 3   **B.** 2 and 3
    C. 2   **D.** 1

6. What is the LCM of 2 and 6?

    A. 2   **B.** 2×2
    C. 2×3 **D.** 12

7. What is the greatest common factor of 9 and 21?

    A. 2   B. 3
    C. 5   D. 7

8. Which of the following is the prime factorization of 14?

    A. $2 \times 2 \times 7$ B. 2×7
    C. 14×1 D. $2^2 \times 7$

9. What is the LCD of $\frac{1}{8}$ and $\frac{1}{4}$?

    A. 8   B. 12
    C. 32   D. 24

10. What are the common factors of 9 and 15?

    A. 3   B. 5
    C. 3 and 5   D. 3, 5, 9, and 15

11. Which of the following is the prime factorization for 12?

    A. $2^3 \times 3$ B. $2^2 \times 3$
    C. 2×6 D. $2(2 \times 3)$

12. What is the greatest common factor of 18 and 36?

    A. 2   B. 3
    C. 4   D. 6

13. Find the LCM of 5 and 8.

    A. 10   B. 20
    C. 30   D. 40

14. Which of the following is the prime factorization of 144?

    A. $2^4 \times 9$ B. $2^4 \times 3^2$
    C. $4^2 \times 9$ D. $2(8 \times 9)$

15. What is the LCD of $\frac{2}{3}$ and $\frac{1}{8}$?

    A. 12   **B.** 15

    C. 20   **D.** 24

16. What is the value of Δ in the expression?

$$\frac{5}{6} = \frac{15}{\Delta}$$

    A. 12   **B.** 15

    C. 18   **D.** 24

17. Which of the following is the equivalent of $\frac{2}{3}$?

    A. $\frac{8}{16}$ **B.** $\frac{3}{9}$

    C. $\frac{3}{6}$ **D.** $\frac{8}{12}$

18. Which of the following is NOT the equivalent of 1.50?

    A. $1\frac{8}{16}$ **B.** $\frac{6}{4}$

    C. $1\frac{5}{10}$ **D.** $1\frac{12}{240}$

19. Which of the following is correct?

    A. $\frac{3}{5} < \frac{3}{50}$ **B.** $\frac{5}{3} > \frac{3}{5}$

    C. $\frac{3}{5}$ is less than $\frac{1}{5}$ **D.** $\frac{3}{5}$ is greater than 1

20. Which of the following is correct?

    A. $\frac{6}{7} \neq$ six out of seven   **B.** $\frac{1}{10}$ is greater than $\frac{3}{100}$.

    C. $\frac{8}{100} = 0.008$ **D.** 2.49 equals to $2\frac{49}{100}$.

21. Which of the following is the equivalent of $1\frac{2}{5}$?

    A. $1\frac{2}{15}$ **B.** $1\frac{6}{15}$

    C. $\frac{7}{15}$ **D.** $\frac{5}{7}$

22. Which of the following is correct?

    A. $\frac{3}{2} < 1\frac{1}{2}$ **B.** $\frac{2}{6} > \frac{1}{3}$

    C. $\frac{4}{10}$ is less than $\frac{2}{5}$ **D.** $1\frac{3}{100}$ is greater than 1

* Use the number line for Exercises **23-25**.

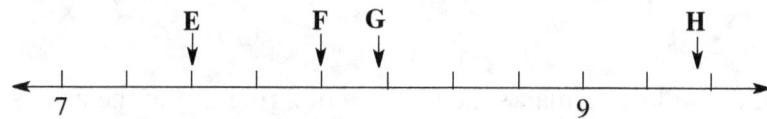

23. What is the relative position of **E**?

    A. $7\frac{1}{2}$ **B.** $7\frac{3}{4}$

    C. 8   **D.** $8\frac{8}{12}$

24. Which of the following is the relative position of $8\frac{3}{16}$?

    A. E   **B.** F

    C. G   **D.** H

25. Which of the following letters are 9 rounded to the nearest tenths?

    A. E, F, G, and H   **B.** F, G, and H

    C. G and H   **D.** H

* Use the number line for Exercises **26-28**.

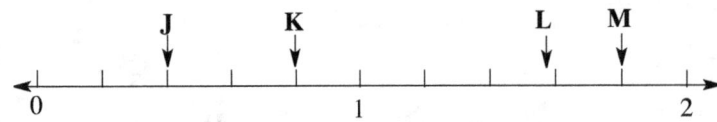

26. What is the relative position of **M**?

A. $\dfrac{12}{15}$ B. $1\dfrac{12}{15}$

C. $\dfrac{3}{5}$ D. $1\dfrac{10}{12}$

27. Which of the following marks the relative position of $\dfrac{4}{5}$?

A. J   B. K
C. L   D. M

28. Which of the following marks the position of 2 rounded to the nearest tenths?

A. J and K   B. K and L
C. L and M   D. M

29. Which of the following lists the numbers from least to greatest?
$$\frac{1}{2}, \ \frac{1}{4}, \ \frac{1}{3}, \ \frac{1}{6}, \ \frac{1}{5}$$

A. $\dfrac{1}{2} < \dfrac{1}{3} < \dfrac{1}{4} < \dfrac{1}{5} < \dfrac{1}{6}$ B. $\dfrac{1}{4} < \dfrac{1}{5} < \dfrac{1}{3} < \dfrac{1}{2} < \dfrac{1}{6}$

C. $\dfrac{1}{5} < \dfrac{1}{6} < \dfrac{1}{3} < \dfrac{1}{4} < \dfrac{1}{2}$ D. $\dfrac{1}{6} < \dfrac{1}{5} < \dfrac{1}{4} < \dfrac{1}{3} < \dfrac{1}{2}$

30. Which of the following lists the numbers from least to greatest?
$$1, 0.4, \ \frac{3}{4}, \ \frac{1}{2}, \ 0.6$$

A. $0.4 < 0.6 < \dfrac{1}{2} < \dfrac{3}{4} < 1$ B. $0.4 < \dfrac{1}{2} < \dfrac{3}{4} < 0.6 < 1$

C. $0.4 < \dfrac{1}{2} < 0.6 < \dfrac{3}{4} < 1$ D. $0.4 < \dfrac{3}{4} < 0.6 < \dfrac{1}{2} < 1$

7. Adding and Subtracting Fractions
2–19. Adding Fractions

$$\frac{1}{5} + \frac{2}{5}$$

SOLUTION

You can add fractions with **like denominators** by adding the numerators.

Add the numerators like whole numbers

$$\frac{1}{5} + \frac{2}{5} = \frac{1+2}{5} = \frac{3}{5}$$

Stay the same denominators

like denominators;
the same numbers of
denominators

So, the sum of $\frac{1}{5} + \frac{2}{5}$ is $\frac{3}{5}$.

2–20. Add the values.

$$1 + \frac{1}{3}$$

SOLUTION

You should rewrite the number as a fraction and then add.

Convert.

$$1 + \frac{1}{3} = \frac{3}{3} + \frac{1}{3} = \frac{4}{3} \text{ or } 1\frac{1}{3}$$

Simply add them.

$$1 + \frac{1}{3} = 1\frac{1}{3}$$

So, the sum of $1 + \frac{1}{3}$ is $1\frac{1}{3}$.

2–21. Find the value of Δ.

$$\frac{1}{4} + \Delta = \frac{3}{4}$$

SOLUTION

For solving Δ, you should subtract $\frac{1}{4}$ from both sides.

$$\frac{1}{4} - \frac{1}{4} + \Delta = \frac{3}{4} - \frac{1}{4}$$

The left side of the numbers can cancel out ($\frac{1}{4} - \frac{1}{4} = 0$) and then only Δ remains.

The right side of the numbers can subtract $\frac{3}{4} - \frac{1}{4} = \frac{2}{4}$ or $\frac{1}{2}$. So Δ is $\frac{1}{2}$.

Exercises 35   Find the sum of the fractions

1) $\dfrac{3}{9} + \dfrac{7}{9}$

2) $\dfrac{1}{7} + \dfrac{6}{7}$

3) $\dfrac{1}{2} + \dfrac{1}{6}$

4) $\dfrac{1}{2} + \dfrac{1}{10}$

5) $\dfrac{1}{3} + \dfrac{1}{6}$

6) $\dfrac{2}{5} + 1$

7) $\dfrac{3}{5} + 2$

8) $1 + 1\dfrac{1}{2}$

9) $\dfrac{3}{4} + \dfrac{1}{8}$

10) $\dfrac{1}{4} + \dfrac{1}{12}$

11) $\dfrac{1}{2} + \dfrac{1}{4}$

12) $\dfrac{1}{2} + \dfrac{1}{8}$

Exercises 36   Find each value of Δ.

1) $\dfrac{1}{2} + \Delta = 1\dfrac{1}{2}$

2) $\dfrac{2}{5} + \Delta = 1$

3) $\Delta + \dfrac{5}{6} = 1\dfrac{1}{3}$

4) $\Delta + 1 = 1\dfrac{2}{3}$

5) $\dfrac{1}{3} + \Delta = 1\dfrac{2}{3}$

6) $\Delta + \dfrac{3}{8} = \dfrac{1}{2}$

Exercises 37   Find the sum of the fractions.

1) $1\frac{1}{5} + \frac{1}{5}$

2) $\frac{1}{3} + 3$

3) $\frac{1}{4} + 1\frac{1}{4}$

4) $3 + 2\frac{1}{2}$

5) $\frac{2}{7} + \frac{5}{7}$

6) $2\frac{1}{3} + 1\frac{2}{3}$

7) $1\frac{1}{2} + \frac{1}{4}$

8) $2\frac{1}{3} + 3$

9) $1\frac{1}{5} + 1\frac{1}{10}$

10) $4 + 1\frac{1}{5}$

Exercises 38   Find each value of Δ.

1) $\frac{5}{6} + \Delta = 1\frac{1}{3}$

2) $1\frac{1}{4} + \Delta = 2$

3) $\Delta + 2 = 3\frac{4}{5}$

4) $\Delta + 1\frac{1}{2} = 3$

5) $\frac{2}{9} + \Delta = \frac{1}{3}$

6) $\Delta + \frac{1}{6} = \frac{2}{3}$

7) $\frac{3}{5} + \Delta = 2 - \frac{3}{5}$

8) $\frac{1}{4} + \Delta = \frac{1}{2} + \frac{3}{4}$

2–22. Subtracting fractions

$$\frac{2}{3} - \frac{1}{3}$$

SOLUTION

You can subtract fractions with **like denominators** by subtracting the numerators.

Subtract the numerators like whole numbers.

$$\frac{2}{3} - \frac{1}{3} = \frac{2-1}{3} = \frac{1}{3}$$

Stay the same denominators.

like denominators;
the same denominators

So, the difference of $\frac{2}{3} - \frac{1}{3}$ is $\frac{1}{3}$.

2–23. Subtract the values.

$$1 - \frac{1}{3}$$

SOLUTION

You should rewrite the whole number as a fraction and then subtract.

Convert.

$$1 - \frac{1}{3} = \frac{3}{3} - \frac{1}{3} = \frac{2}{3}$$

like denominators;
the same denominators

So, the difference of $1 - \frac{1}{3}$ is $\frac{2}{3}$.

2–24. Find the value of Δ.

$$\Delta - \frac{1}{4} = \frac{3}{4}$$

SOLUTION

For solving Δ, you should add $\frac{1}{4}$ to both sides.

$$\Delta - \frac{1}{4} + \frac{1}{4} = \frac{3}{4} + \frac{1}{4}$$

The left side of the numbers can be canceled out ($\frac{1}{4} - \frac{1}{4} = 0$) and then only Δ

remains. The right side of the numbers can be added $\frac{3}{4} + \frac{1}{4} = \frac{4}{4}$ or 1. So Δ is 1.

Exercises 39   Find the difference of the fractions.

1) $\dfrac{5}{8} - \dfrac{1}{8}$

2) $\dfrac{4}{6} - \dfrac{1}{6}$

3) $\dfrac{3}{8} - \dfrac{1}{8}$

4) $1 - \dfrac{1}{4}$

5) $1 - \dfrac{7}{9}$

6) $1\dfrac{1}{2} - 1\dfrac{3}{6}$

7) $1\dfrac{1}{3} - 1$

8) $1\dfrac{7}{8} - 1$

9) $\dfrac{9}{7} - \dfrac{1}{7}$

10) $\dfrac{6}{5} - \dfrac{3}{5}$

Exercises 40   Find each value of Δ.

1) $\dfrac{1}{2} - \Delta = \dfrac{1}{4}$

2) $\dfrac{5}{8} - \Delta = \dfrac{1}{4}$

3) $\Delta - \dfrac{1}{2} = \dfrac{1}{2}$

4) $\Delta - \dfrac{3}{5} = \dfrac{2}{5}$

5) $1\dfrac{1}{6} - \Delta = \dfrac{2}{3}$

6) $\Delta - \dfrac{1}{3} = 1\dfrac{2}{3}$

7) $2 - \Delta = \dfrac{7}{9}$

8) $\Delta - 2 = 1\dfrac{2}{3}$

Exercises 41   Find the difference of the fractions.

1) $3 - \dfrac{1}{4}$

2) $5\dfrac{7}{8} - 3$

3) $2\dfrac{1}{3} - 1$

4) $3\dfrac{1}{3} - 3$

5) $3\dfrac{9}{7} - 1\dfrac{1}{7}$

6) $3\dfrac{4}{5} - 2\dfrac{4}{5}$

7) $\dfrac{1}{2} - \dfrac{1}{4}$

8) $1\dfrac{1}{3} - \dfrac{1}{6}$

9) $\dfrac{2}{3} - \dfrac{2}{6}$

10) $\dfrac{10}{15} - \dfrac{1}{3}$

11) $\dfrac{18}{21} - \dfrac{4}{7}$

12) $\dfrac{9}{12} - \dfrac{1}{4}$

Exercises 42   Find each value of Δ.

1) $\dfrac{5}{6} - \Delta = \dfrac{1}{3}$

2) $\Delta - \dfrac{3}{4} = \dfrac{1}{4}$

3) $1\dfrac{1}{5} - \Delta = \dfrac{3}{5}$

4) $1\dfrac{1}{3} - \Delta = \dfrac{1}{3}$

5) $1\dfrac{5}{8} - \Delta = 1\dfrac{1}{4} + \dfrac{1}{4}$

6) $1\dfrac{4}{9} - \Delta = \dfrac{2}{3} + \dfrac{4}{9}$

* Solving Problems

Exercises 43   Solve each problem using the given information.

For Exercises **1-5**. Mrs. Chan is preparing to make pizza for Thanksgiving. The ingredients are 16.8 ounces of tomato sauce, $2\frac{1}{2}$ cups of shredded mozzarella cheese, 27.6 ounces of pizza crust dough, $1\frac{1}{8}$ cups of chopped pepperoni slices, and $\frac{1}{2}$ teaspoons of garlic powder. This recipe can serve 24 people.

1) What is the sum of the ounces of tomato sauce and pizza crust dough?

2) What is the difference between the cups of shredded mozzarella cheese and chopped pepperoni slices?

3) How much cups of chopped pepperoni slices does she need to make in order to serve 12 people?

4) How many teaspoons of garlic powder does she need to make it in order to serve 18 people?

5) How many ounces of tomato sauce and pizza crust dough does she need to make it in order to serve 6 people?

1. Find the difference of the expression below.

$$\frac{3}{5} - \frac{1}{5}$$

    A. $\frac{2}{5}$ **B.** $\frac{1}{10}$

    C. $\frac{1}{5}$ **D.** $\frac{3}{10}$

* Use the information for Exercises **2-7**. There are 120 M&Ms on your table. There are four colors. $\frac{4}{16}$ of the M&Ms are red, $\frac{1}{8}$ of the M&Ms are yellow, and $\frac{2}{16}$ of the M&Ms are brown.

2. How many red M&Ms are on the table?

    A. 30   **B.** 40
    C. 45   **D.** 50

3. What fraction is the sum of red and yellow M&Ms?

    A. $\frac{5}{7}$ **B.** $\frac{3}{8}$

    C. $1\frac{1}{28}$ **D.** $\frac{5}{12}$

4. How many yellow M&Ms are on the table?

    A. $7\frac{1}{2}$ **B.** 30
    C. 14   **D.** 15

5. What fraction is the difference between red and brown M&Ms?

    A. $\frac{1}{8}$ **B.** $\frac{1}{16}$

    C. $\frac{3}{16}$ **D.** $\frac{2}{8}$

6. How many piles of M&Ms can you make them if each pile has 15 pieces?

    A. 6   **B.** 8
    C. 10   **D.** 12

7. What fraction represents the number of the remaining color?

     A. $\dfrac{1}{16}$ B. $\dfrac{1}{8}$

     C. $\dfrac{1}{4}$ D. $\dfrac{1}{2}$

8. Which of the following is the sum of $\dfrac{6}{7}+\dfrac{6}{7}$?

     A. $\dfrac{12}{14}$ B. $1\dfrac{5}{7}$

     C. $2\dfrac{6}{7}$ D. $\dfrac{36}{49}$

* Use the information for Exercises **9-10**. At the aquarium, the orcas are fed 50 pounds of fish. The dolphins are fed 3/5 as many pounds of fish as the orcas.

9. How many pounds of fish do the dolphins eat?

     A. 30   **B.** 40
     C. 45   **D.** 50

10. If the seals are fed 3/4 as much as what the orcas are fed, what is the fraction representing the difference of the feeding amount between the seals and the dolphins?

     A. $\dfrac{3}{10}$ B. $\dfrac{9}{20}$

     C. $\dfrac{3}{20}$ D. $\dfrac{6}{10}$

* Use the information for Exercises **11-12**. Rio has $28 in the wallet. He decides to spend $\dfrac{1}{4}$ of his money for lunch and $\dfrac{1}{10}$ for a snack.

11. What fraction represents the sum of the lunch and the snack?

     A. $\dfrac{7}{40}$ B. $\dfrac{7}{20}$

     C. $1\dfrac{7}{10}$ D. $\dfrac{5}{12}$

12. What is the sum of the cost of the lunch and snack?

A. $7.00   B. $8.00

C. $9.80   D. $10.80

13. What fraction represents the difference between the lunch and snack?

A. $\frac{1}{16}$ B. $\frac{1}{4}$

C. $\frac{3}{40}$ D. $\frac{3}{20}$

14. What is the sum of $2\frac{2}{3} + 1$?

A. $3\frac{4}{6}$ B. $3\frac{2}{3}$

C. 3   D. 4

15. What is the value of Δ given in the expression below?
$$\frac{2}{3} - \Delta = \frac{1}{3}$$

A. $\frac{1}{3}$ B. $\frac{3}{5}$

C. $1\frac{1}{6}$ D. $\frac{1}{6}$

16. What is the value of Δ given in the expression below?
$$1\frac{4}{6} - \Delta = \frac{5}{6}$$

A. $\frac{4}{6}$ B. $\frac{5}{6}$

C. $\frac{1}{6}$ D. $\frac{4}{3}$

17. What is the difference of $2 - \frac{4}{9}$?

A. $\frac{5}{9}$ B. $1 + \frac{5}{9}$

C. $1\frac{4}{9}$ D. $1\frac{1}{3}$

8. Relationship between Mixed Numbers
2–25.   Relationship between Mixed Numbers

	Dividing fractions	Dividing decimals	Fractions to decimals
1)	$1 \div 10 = \frac{1}{10}$	$1 \div 10 = 0.1$	$\frac{1}{10} = 0.1$
2)	$1 \div 100 = \frac{1}{100}$	$1 \div 100 = 0.01$	$\frac{1}{100} = 0.01$
3)	$11 \div 10 = 1\frac{1}{10}$	$11 \div 10 = 1.1$	$\frac{11}{10} = 1.1$
4)	$12 \div 10 = 1\frac{2}{10}$	$12 \div 10 = 1.2$	$\frac{12}{10} = 1.2$
5)	$21 \div 10 = 2\frac{1}{10}$	$21 \div 10 = 2.1$	$\frac{21}{10} = 2.1$
6)	$22 \div 10 = 2\frac{2}{10}$	$22 \div 10 = 2.2$	$\frac{22}{10} = 2.2$

Exercises 44   Write each value as a decimal. If necessary, round to the nearest hundredth.

1) $\frac{1}{2}$ _____

2) $\frac{1}{10}$ _____

3) $\frac{3}{8}$ _____

4) $\frac{1}{15}$ _____

5) $\frac{6}{7}$ _____

6) $5\frac{3}{4}$ _____

7) $\frac{2}{3}$ _____

8) $\frac{2}{6}$ _____

9) $\frac{4}{50}$ _____

10) $\frac{6}{24}$ _____

11) $\frac{6}{8}$ _____

12) $2\frac{3}{5}$ _____

Exercises 45   Write each value as a fraction.

1) 0.25

2) 0.7

3) 20.56

4) 5.2

5) 0.339

6) 8.101

7) 0.75

8) 0.007

9) 4.18

10) 1.28

11) 2.30

12) 2.14

Exercises 46   Write each value as either a fraction or decimal

1) 0.8

2) 2.5

3) $\frac{5}{7}$

4) $\frac{1}{25}$

5) 0.3

6) $2\frac{7}{8}$

7) 0.22

8) 2.05

9) $\frac{5}{10}$

10) $\frac{101}{1000}$

11) 10.45

12) $2\frac{7}{100}$

1. Which of the following forms is not 0.35?

    A. $\dfrac{3}{10} + \dfrac{5}{100}$ **B.** thirty-five hundredths

    C. $\dfrac{35}{100}$ **D.** $35 \div 10$

2. Which of the following forms is not $2\dfrac{3}{100}$?

    A. 2.03   **B.** $2 \times (3 \div 100)$
    C. $2 + (3 \div 100)$ **D.** Add the quotient of three divided by hundred.

3. Which of the following forms is 1.001?

    A. $1 + \dfrac{1}{100}$ **B.** $1 \times (1 \div 1000)$

    C. $1 \times \dfrac{1}{100}$ **D.** one and one thousandth

4. Which of the following forms is 0.403?

    A. $\dfrac{4}{10} + \dfrac{3}{1000}$ **B.** four hundred three thousands

    C. $\dfrac{403}{100}$ **D.** $0.402 \div 1000$

5. Which of the following forms is not $\dfrac{3}{7} + \dfrac{3}{5}$?

    A. $(3 \div 7) + (3 \div 5)$
    B. Add the quotient of three and seven with the quotient of three and five.
    C. $\dfrac{3}{7} + \dfrac{3}{5} = \dfrac{15}{35} + \dfrac{21}{35}$
    D. $(3 \times 7) \div (3 \times 5)$

6. Which of the following forms is $\dfrac{1}{4}$?

    A. $4\overline{)1}$ **B.** $4\overline{)0.1}$
    C. $1\overline{)4}$ **D.** $0.1\overline{)0.4}$

7. Which of the following can be 2.06?

    A. $\dfrac{206}{1000}$

    B. $\dfrac{6}{200}$

    C. $2\dfrac{6}{1000}$

    D. $2\dfrac{6}{100}$

8. Which of the following can be $2\dfrac{2}{5}$?

    A. 2.04

    B. $2 + 0.4$

    C. 0.24

    D. 2×0.4

9. Which of the following can be 53.103?

    A. $\dfrac{53}{1000} + \dfrac{103}{1000}$

    B. $53\dfrac{103}{1000}$

    C. $53\dfrac{1}{103}$

    D. $53\dfrac{103}{100}$

10. Which of the following can be $1\dfrac{5}{8}$?

    A. 106.25

    B. 1.0625

    C. 10.625

    D. 10.58

11. Which of the following is the sum of $0.6 + \dfrac{3}{50}$?

    A. 0.66

    B. 0.606

    C. 6.06

    D. 6.006

12. Which of the following is the difference of $0.525 - \dfrac{6}{15}$?

    A. 0.125

    B. 1.25

    C. 0.0125

    D. 1.025

13. Which of the following is the product of $1.06 \times \dfrac{2}{5}$?

    A. 4.24

    B. 0.424

    C. 1.46

    D. 0.146

CHAPTER 3
Patterns, Function, and Algebra

In this chapter, you will identify number sense based on the number system, algebraic expression, and order of operations. Also you will learn about graphing functions, plotting the coordinate, relating function tables, and finding the equations of function tables.

1. Order of Operations

3–1. Order of operations

> a) When solving a problem with more than one operations (+, /, *, -), there is a standard order of operations. First, perform any calculations inside the parentheses. Next perform all multiplications and divisions, working from left to right. Finally, perform all additions and subtractions, working from left to right.
>
> b) In the absence of any parentheses or grouping symbols, multiplication and division is always done in order from left to right before addition and subtraction.

Exercises 1   Solve each expression.

1) $2 \times (3 + 2)$ _____

2) $2 + (4 \div 2) - 1$ _____

3) $(6 \div 2) - (3 \times 2)$ _____

4) $(3 \times 2) + 2$ _____

5) $(3 - 1) \times (4 - 2)$ _____

Exercises 2   Write each word form in its standard form and solve its expression.

1) The quotient of four and two decreased by two _____

2) Multiply the sum of three and six by two, then add three. _____

3) Multiply the difference of four and one by five _____

4) Add the sum of five and three with the quotient of fifteen and five.

Exercises 3   Solve each expression.

1) $12 + (84 + 11)$

2) $(23 - 9) + 28$

3) $(12 - 6) + (20 - 7)$

4) $42 - (31 - 9) + 30$

5) $8 + 2 \times 2$

6) $4 \times 3 - 4$

7) $5 + 7 - 3 \times 4$

8) $24 \div 6 - 2$

9) $14 - 24 \div 4$

10) $8 + 4 \div 2$

11) $42 + (49 + 21)$

12) $(63 - 9) + 58$

Exercises 4   What operation sign or value is Δ if the equation is true?

1) $9 \div 3 + 3 = 3 \,\Delta\, 3$

2) $4 \,\Delta\, 3 - 3 = 3 \times 3$

3) $5 - 2 \div 2 = (\Delta - 2) \times 2$

4) $27 \div 3 + 3 = (\Delta + 3) \times 3$

5) $3 + 3 \times 3 = \Delta \times 4$

6) $3 \times 3 - 3 = 3 + \Delta \div 3$

7) $(2 + \Delta) + 2 = 3 \times 2$

8) $\Delta \div 3 = 3 \times 3$

Exercises 5   Solve each expression.

1) $13 - (3 \times 4) + 2$ **2)** $(15 - 10) \times (3 + 2)$

_____   _____

3) $20 \div 4 - 4 \times 2 + 8$ **4)** $[(25 \div 5) \times (6 - 3)] \div 2$

_____   _____

5) $6 - (2 \times 2) \div 2$ **6)** $[5 - (4 \div 4)] + [(2 \times 3) + 3]$

_____   _____

7) $4 \times 2 + 4 \div 2$ **8)** $12 \div 3 - 3 \times (3 - 3)$

_____   _____

9) $2 \times 3 - 2 \times 3$ **10)** $[(15 \div 5) \times (6 - 3)] \div 3 - 2$

_____   _____

11) $[5 - (8 \div 4)] + [(2 \times 3) + 6]$ **12)** $4 - (2 \div 2) \times 2$

_____   _____

Exercises 6   What operation sign or value is Δ if the equation is true?

1) $(9 + 4) \div 2 = \Delta$ **2)** $12 \div (3 \times 2) = \Delta$

_____   _____

3) $9 + 9 + 9 = \Delta \times 9$ **4)** $(20 - 4) \times 2 = \Delta + 16$

_____   _____

5) $10 \times 2 = \Delta + 10$ **6)** $8 \div (\Delta + 2) = 2$

_____   _____

7) $24 \div \Delta = 24 - 12$ **8)** $5^{\Delta} = 5 \times 5 \times 5 \times 5 \times 5$

_____   _____

9) $5 \times 5 + 5 = (1 \, \Delta \, 5) \times 5$ **10)** $\Delta^2 = (2 \times 2)^2$

_____   _____

* Solving Problems

Exercises 7   Solve each problem using the given information.

1) Lucy would like to sell 120 cups of lemonade for the school fundraiser. The cups come in 3 different colors and have an equal number of each. She sells 35 yellow cups, 30 blue cups, and 38 red cups. How many cups does she have left over in total? How many blue cups are left over?

2) At the school festival, 1,062 out of 1,255 hot dogs are eaten. How many hot dogs are left?

3) McKenzie puts different amounts of candy in 3 jars. The first jar contains 4 pieces of candy. The second jar has half the amount of candy in the first jar. The third jar contains twice as many candy than the first and second jar combined. How many pieces of candy does McKenzie have?

4) At swim practice, the swimmers are swimming laps. If there are 11 swimmers and they swim a total of 132 laps, how many laps did each person swim?

5) Find the value of the expression below.
    Subtract the quotient of nine and three from the product of fourteen and one.

6) Find the value of the expression below.
    Multiply the product of three and two to the sum of five and three.

7) Find the value of the expression below.
    Add the quotient of eight and two with the difference of fifteen and seven.

SELF-TEST

1. Which of the following is the value of the expression below?
    Eleven times the sum of one and six

    A. 88   B. 77
    C. 66   D. 55

2. Which of the following best describes the expression below in standard form?
    Add the quotient of ten and five with the product of three and two

    A. $(10 + 2) \div (3 \times 2)$ B. $(10 \times 2) \div (3 + 2)$
    C. $(10 \div 2) + (3 \times 2)$ D. $(10 \div 2) \times (3 + 2)$

3. Which of the following best describes the expression below in word form?
    $(9 + 6) \div (3 \times 2)$

    A. the quotient of the sum of nine and six and the difference of three and two
    B. the sum of nine and six divided by the product of three and two
    C. the sum of nine and six divided by three decreased by two
    D. add nine and six divided by three increased by two

4. What is the value of the expression below?
    Subtract the quotient of thirty-eight and two with the product of ten and one.

    A. 8   B. 9
    C. 10   D. 29

5. What is the value of the expression below?
    $(60 - 12) \div 6 + 6$

    A. 6   B. 8
    C. 4   D. 14

6. What is the value of the expression below?
    $3 + 7 \times 2 - 6 \div 2$

    A. 17   B. −10
    C. 14   D. −20

7. Find the operation of Δ.
$$1 \, \Delta \, 7 \times 7 = (7 + 7) \times 4$$

A. ÷

B. ×

C. +

D. −

8. Find the operation that represents Δ.
$$14 \, \Delta \, 7 + (14 - 6) \div 2 = 11$$

A. ÷

B. ×

C. +

D. −

9. Find the value that represents Δ.
$$\Delta \times 3 = 10 + 10 + 10$$

A. 30

B. 20

C. 10

D. 3

10. Find the operation that represents Δ.
$$50 \, \Delta \, 10 = 10 + 10 + 10 + 10$$

A. ÷

B. ×

C. +

D. −

11. What is the value of the expression below?
$$18 \times 2 - 18 \div 2$$

A. −4.5

B. 35

C. 27

D. 9

12. James is driving his car 45 miles per hour while Kim is driving her car 67 miles per hour. How much faster is Kim going than James?

A. 112 miles per hour

B. 22 miles per hour

C. 1.5 miles per hour

D. 12 miles per hour

13. The population of Elizabeth, New York is 125,660 people while the population of Anaheim, California is 341,361 people. How many more people does Anaheim have than Elizabeth?

A. 467,021   **B.** 215,701
C. 2.7   **D.** 216,701

14. A miser is organizing his coins into several bags. He has organized his coins so that there are 32 coins in each bag. If the miser has 224 coins in total, how many bags of coins does he have?

    A. 7 bags   **B.** 192 bags
    C. 256 bags   **D.** 7,168 bags

15. Reorder the expression below so that it follows the order of operations
$$61 \times 18 - 25 \div 5 + 74$$

    A. $61 \times (18 - 25) \div 5 + 74$ **B.** $61 \times (18 - 25) \div (5 + 74)$
    C. $(61 \times 18) - (25 \div 5) + 74$ **D.** $(61 \times 18) - 25 \div (5 + 74)$

16. Find the value of the expression below.
$$4 + 3 \times 6 - 4 \div 2$$

    A. 19   **B.** 40
    C. 20   **D.** 54

17. At a fundraiser, 11 people each donated the same amount of money. If the money they donated adds up to $187, how much did each person donate?

    A. $2,057   **B.** $17
    C. $176   **D.** $198

18. Reorder the expression below so that it follows the order of operations.
$$24 \times 42 \div 8 - 3$$

    A. $24 \times [(42 \div 8) - 3]$ **B.** $[24 \times (42 \div 8)] - 3$
    C. $[(24 \times 42) \div 8] - 3$ **D.** $(24 \times 42) \div (8 - 3)$

2. Solving Equations with One Variable: Addition and Subtraction

3-2. Find the value of the variable.

$$4 + x = 15$$

> **SOLUTION**
>
> i) First, look at the expression in the number.
> $4 + x = 15$, where x is a variable that represents the unknown value.
>
> $\underbrace{4 + x}_{\text{Left}} = \underbrace{15}_{\text{Right}}$ An **equation** is considered correct when both sides of the equation are equal.
>
>   So the left side $(4 + x)$ has the same value as the right side (15).
>
> ii) Now you should find the value of x. For solving x, you should subtract 4 from both sides.
> $4 - 4 + x = 15 - 4$
>   The left side of the numbers can be canceled out $(4 - 4 = 0)$ and then only x remains. The right side of the numbers can be subtracted $(15 - 4 = 11)$.
>   So x is 11.
>
> Check yourself
> Check if x is 11.
> $4 + x = 15$
>   Replace x with 11.
> $4 + 11 = 15$
> $15 = 15$
> So both sides are equal. It means true that x is 11.

3-3. Find the value of the variable.

$$y + 9 = 24$$

> **SOLUTION**
>
> For solving y, you should subtract 9 from both sides and then only y will remain on the left side.
> $y + 9 = 24$ Original equation
> $y + 9 - 9 = 24 - 9$ Subtract 9 from both sides.
> $y = 15$ Simplify. $9 - 9 = 0$, $24 - 9 = 15$
> So y is 15.
>
> Check whether y is 15 or not.
> $y + 9 = 24$ Original equation
> $15 + 9 = 24$ Replace y with 15.
> $24 = 24$ So $y = 15$ is true.

3-4.   Find the value of $\square$.

$$7 + \square = 34$$

SOLUTION

An equation with $\square$ is similar to solving a variable.

$7 + \square = 34$ Original equation

$7 - 7 + \square = 34 - 7$ Subtract 7 from both sides.

$\square = 27$ Simplify. $7 - 7 = 0$, $34 - 7 = 27$

So the value of $\square$ is 27.

Check whether the value is 27 or not.

$7 + \square = 34$ Original equation

$7 + 27 = 34$ Replace $\square$ with 27.

$34 = 34$ It means that $\square = 27$ is true.

Exercises 8   Solve each equation.

1) $6 + x = 18$

2) $n + 9 = 36$

3) $x + 2 = 44$

4) $9 + y = 29$

5) $5 + x = 13$

6) $n + 4 = 18$

7) $n + 8 = 25$

8) $7 + y = 54$

9) $4 + x = 64$

10) $n + 10 = 36$

11) $x + 12 = 56$

12) $15 + y = 83$

Exercises 9   Solve each equation.

1) $83 + x = 118$

2) $n + 98 = 306$

3) $x + 72 = 124$

4) $38 + y = 99$

5) $105 + x = 132$

6) $n + 49 = 108$

7)

$$\begin{array}{r} 45 \\ + \ y \\ \hline 83 \end{array}$$

8)

$$\begin{array}{r} x \\ + \ 8 \\ \hline 89 \end{array}$$

9)

$$\begin{array}{r} 61 \\ + \ n \\ \hline 179 \end{array}$$

10)

$$\begin{array}{r} 21 \\ + \ n \\ \hline 43 \end{array}$$

11)

$$\begin{array}{r} 61 \\ + \ y \\ \hline 86 \end{array}$$

12)

$$\begin{array}{r} x \\ + \ 7 \\ \hline 21 \end{array}$$

Exercises 10   Solve each expression.

1)   If $x = 12$, find the value of $8 + x - 5$.

2)   If $x = 5$, find the value of $x - 3$.

3)   If $x = 2$, find the value of $8 \times 2 + x$.

4)   If $c = 18$, find the value of $(c + 8) \div 2$.

5)   If $y - 1 = 1$, find the value of $y + 15$.

6)   If $x + 4 = 7$, find the value of $(4 + x) - 1$.

Exercises 11   Solve each equation.

1) $n + 71 = 125$

2) $77 + y = 154$

3) $36 + \Delta = 88$

4) $n + 69 = 114$

5) $x + 94 = 144$

6) $106 + \Delta = 132$

7)
$$\begin{array}{r} 42 \\ + \square \\ \hline 93 \end{array}$$

8)
$$\begin{array}{r} 125 \\ + \square \\ \hline 186 \end{array}$$

9)
$$\begin{array}{r} \square \\ + 24 \\ \hline 45 \end{array}$$

10)
$$\begin{array}{r} 172 \\ + \square \\ \hline 401 \end{array}$$

11)
$$\begin{array}{r} 101 \\ + \square \\ \hline 246 \end{array}$$

12)
$$\begin{array}{r} \square \\ + 67 \\ \hline 105 \end{array}$$

Exercises 12   Solve each expression.

1)   If $2 - x = 1$, find the value of $8 + x$.

2)   If $2 + c = 5$, find the value of $c + 8$.

3)   If $y - 3 = 4$, find the value of $y + 5$.

4)   If $x + 3 = 4$, find the value of $5 + x$.

5)   If $5 - x = 2$, find the value of $8 - (x + 2)$.

6)   If $6 + x = 7$, find the value of $x + 6 + 7$.

3-5.   Find the value of each variable.

$$17 - x = 8$$

SOLUTION

First, look at the expression.

$17 - x = 8$	Original equation
$17 - 17 - x = 8 - 17$	Subtract **17** from both sides.
$-x = -9$	Simplify. $17 - 17 = 0, 8 - 17 = -9$
$-(-x) = -(-9)$	Multiply (-1) on both sides.
$x = 9$	$-(-x) = x$

So x is 9.

Check yourself
Check whether x is 9 or not.
$$17 - x = 8$$
Replace x with 9.
$$17 - 9 = 8$$
$$8 = 8$$
So it is true that x is 9.

﹡ Important operation system

even	odd
$(-) \times (-) = +$	$(+) \times (-) = -$
$(+) \times (+) = +$	$(-) \times (+) = -$

3-6.   Find the value of each variable.

$$y - 5 = 12$$

SOLUTION

$y - 5 = 12$	Original equation
$y - 5 + 5 = 12 + 5$	Add **5** to both sides.
$y = 17$	Simplify. $-5 + 5 = 0, 12 + 5 = 17$

So the variable y is 17.

Check yourself

$y - 5 = 12$	Original equation
$17 - 5 = 12$	Replace y with 17.
$12 = 12$	So, that means $y = 17$ is true.

3–7. Find the value of $\square$.

$$\square - 15 = 16$$

SOLUTION

An equation with $\square$ is the same thing as solving an equation with a variable.

$\square - 15 = 16$	Original equation
$\square - 15 + \mathbf{15} = 16 + \mathbf{15}$	Add **15** to both sides.
$\square = 31$	Simplify. 15 - 15 = 0, 16 + 15 = 31

Check yourself

$\square - 15 = 16$	Original equation
$31 - 15 = 16$	Replace $\square$ with 31.
$16 = 16$	It means that $\square = 31$ is true.

Exercises 13   Find the value of the variable in each equation.

1) $a - 3 = 2 + 11 - 8$

2) $3 - 18 = 19 - x$

3) $1 + a = 4 - 2 + 1$

4) $a - 12 = 3 - 11$

5) $72 + 28 - 23 = 14 + k - 15$

6) $3 + a = 19 + 2 + 11$

7) $20 + c - 7 = 92 - 12 + 9$

8) $n - (71 - 15) = 13 - 2$

9) $(x - 13) + 51 = 77$

10) $8 + 9 - 13 = 2 - 12 + n$

11) $39 - 13 = 20 - (4 - x)$

12) $2 - 3 = n - 5$

Exercises 14   Solve each equation.

1) $n - 7 = 23$

2) $28 - y = 12$

3) $17 - x = 9$

4) $n - 14 = 25$

5) $c - 7 = 15$

6) $15 - y = 8$

7)
$$\begin{array}{r} 52 \\ -\ n \\ \hline 23 \end{array}$$

8)
$$\begin{array}{r} 105 \\ -\ y \\ \hline 86 \end{array}$$

9)
$$\begin{array}{r} x \\ -\ 17 \\ \hline 45 \end{array}$$

10)
$$\begin{array}{r} 86 \\ -\ y \\ \hline 48 \end{array}$$

11)
$$\begin{array}{r} x \\ -\ 32 \\ \hline 63 \end{array}$$

12)
$$\begin{array}{r} 69 \\ -\ n \\ \hline 15 \end{array}$$

Exercises 15   Find the value of each expression.

1)   If $n = 6$, find the value of $8 - n$.

2)   If $c = 12$, find the value of $c - 8$.

3)   If $x - 28 = 19$, find the value of $42 - x + 7$.

4)   If $4 - y = 16$, find the value of $15 - y$.

5)   If $x = 10$, find the value of $x - (1 + 5)$.

6)   If $x = 8$, find the value of $18 - (x - 5)$.

Exercises 16   Solve the equation.

1) $\Delta - 9 = 23$ 2) $78 - y = 41$

_____   _____

3) $81 - \Delta = 39$ 4) $28 - \Delta = 21$

_____   _____

5) $\Delta - 60 = 27$ 6) $102 - y = 56$

_____   _____

Exercises 17   Find the value of each expression.

1)   If $1 + n = 6$, find the value of $6 - n$.   2)   If $1 - c = 4$, find the value of $1 - c - 2$.

_____   _____

3)   If $x - 15 = 4$, find the value of $23 - x$.   4)   If $12 - y = 29$, find the value of $15 - y$.

_____   _____

5)   If $x - 6 = 10$, find the value of $x - (1 + 5)$.   6)   If $x - 5 = 3$, find the value of $18 - (x - 5)$.

_____   _____

Exercises 18   Find the value of each variable.

1) $18 - x = 14 - 8$ 2) $2 - 5 = n - (4 + 2)$

_____   _____

3) $14 - 12 = s - 6$ 4) $82 - r = 28 + 17$

_____   _____

5) $35 + a = 77 - 23$ 6) $64 - 51 = s - 16$

_____   _____

SELF-TEST

1. Which of the following expressions represents "the sum of 18 and x"?

    A. $18 \div x$ **B.** $18 - x$
    C. $18 + x$ **D.** $18 = x$

2. What is the value of x for the equation below?
$$29 + x = 53$$

    A. 24   **B.** −24
    C. 34   **D.** 92

*   For Exercises **3-4**. McKenzie puts different amounts of candy in 3 jars. The first jar contains 4 pieces of candy. The second jar has three times of the amount of candy in the first jar. He does not remember how many pieces was put into the third jar.
3. What is the expression of the problem if she has 28 total pieces of candy?

    A. $(k + 4) + 3 = 28$ **B.** $4 + (3 \times k) = 28$
    C. $(k \times 4) + (3 \times k) = 28$ **D.** $(k + 4) + (3 \times k) = 28$

4. How many pieces of candy would she have in total? Use the expression from the previous problem.

    A. 2   **B.** 4
    C. 6   **D.** 8

*   For Exercises **5-6**. The ice cream store sells vanilla, strawberry, and chocolate flavors. On Saturday, Taylor sold 14 vanilla ice cream cones and 13 strawberry ice cream cones.
5. Find the equation that represents how many chocolate ice cream cones were sold if she had sold 76 ice cream cones in total?

    A. $76 + 14 + 13 = k$ **B.** $14 + 13 + k = 76$
    C. $(14 + 13) \times k = 76$ **D.** $76 = k - 14 - 13$

6. How many chocolate ice cream cones does she sell?

    A. 26   **B.** 27
    C. 50   **D.** 49

* For Exercises **7-10**. Lisa spent 7 hours in the garden. She planted some seeds and watered the plants for an hour, and then spends 3 hours weeding the garden. For the rest of the time, she trimmed the garden.

7. What is the equation used to find how long Lisa trimmed her garden?

    A. $7 = x + 1 + 3$ **B.** $7 + x = 1 + 1 + 3$

    C. $k = 7 + 1 + 1 + 3$ **D.** $7k = 1 + 1 + 3$

8. How many hours did she spend trimming the garden?

    A. 1 hour   **B.** 2 hours

    C. 3 hours   **D.** 4 hours

9. Which of the following expressions represents "the difference of 18 and x"?

    A. $18 \div x$ **B.** $18 - x$

    C. $18 + x$ **D.** $18 = x$

10. What is the value of x for the equation below?
$$x - 54 = 17$$

    A. 34   **B.** 3

    C. 71   **D.** 918

* For Exercises **11-12**. An electronic store has 504 laptops and desktops. So far, the owner has sold 154 laptops and 17 desktops in a week, but four customers have each returned a desktop.

11. What is the equation that represents this problem?

    A. $504 = x + 154 + 17 - 4$ **B.** $504 - 4 = 154 + 17 - x$

    C. $504 - 4 = x + 154 + 17$ **D.** $504 - 154 - 17 = x + 4$

12. How many computers are left in the store?

    A. 356   **B.** 354

    C. 335   **D.** 337

13. Lisa has \$32.00 in her purse. She earns \$18.00 more dollars through babysitting, but

she spent $5.00 for an ice cream. Which equation could be used to show how much money she has now?

A. $32 - 5 = x + 18$ 　　　　　　　**B.** $32 + 18 = x + 5$
C. $32 - 18 = x + 5$ 　　　　　　　**D.** $18 - 5 = x + 32$

14. How much money does she have in her purse?

A. $9.00 　　　　　　　　　　　　**B.** $45.00
C. $55.00 　　　　　　　　　　　**D.** $19.00

* For Exercises **15-16**. There are 49 pieces of peppermint candy in the bowl. Kelly ate 2 pieces of candy and gave some pieces of candy to her friend. By the end of the day, there are 23 pieces left.
15. Which equation expresses the problem?

A. $49 - 2 + x = 23$ 　　　　　　　**B.** $23 + x = 49 - 5$
C. $49 = 2 + 23 + x$ 　　　　　　　**D.** $23 - 2 = x + 49$

16. How many pieces of candy did she give to her friend?

A. 22 　　　　　　　　　　　　　**B.** 24
C. 26 　　　　　　　　　　　　　**D.** 28

* For Exercises **17-18**. Nick borrowed 12 books from the library. He read 5 fiction books and returned 4 of them.
17. What is the equation showing how many fiction books he has?

A. $12 = 4 + x$ 　　　　　　　　　**B.** $12 \times x = 5 + 4$
C. $12 = x - 5 + 4$ 　　　　　　　**D.** $5 - 4 = x \times 12$

18. How many fiction books does he have?

A. 6 　　　　　　　　　　　　　　**B.** 7
C. 8 　　　　　　　　　　　　　　**D.** 9

3. Solving Equations with One Variable: Multiplication and Division

3–8. Solve the equation.

$$7 \times n = 98$$

SOLUTION

First find the value of n. To solve for n, divide each side by 7, which cancels out the 7 on the left side. Therefore, only n and 14 will be left on both sides.

The operation sign ($\times$) can be omitted between a number and a variable without changing the meaning.

$7 \times n = 98$ or $7n = 98$

$\frac{1}{7} \times 7 \times n = 98 \times \frac{1}{7}$ Divide each side by 7.

$\frac{1}{7} \times 7 \times n = 98 \times \frac{1}{7}$ Cancel out 7 on both sides.

$7 \times 1 = 7$
$7 \times 1 = 7$
$7 \times 14 = 98$

$n = 14$ Simplify.

So, the solution of $7 \times n = 98$ is 14.

Check

Also you can check the solution of $n = 14$ by substituting the value in the equation of $7 \times n = 98$.

    $7 \times 14 = 98$ Substitute n with 14.
    $98 = 98$ It means that $n = 14$ is true.

This means you solved the equation correctly.

3–9. Find the value of $\square$.

$$\square \times 6 = 114$$

SOLUTION

An equation with $\square$ is the same thing as solving an equation with a variable.

    $\square \times 6 = 114$ Original equation

    $\square \times 6 \times \frac{1}{6} = 114 \times \frac{1}{6}$ Divide each side by 6.

    $\square \times 6 \times \frac{1}{6} = 114 \times \frac{1}{6}$ Cancel out 6 on both sides.

    $6 \times 1 = 6$
    $6 \times 1 = 6$
    $6 \times 19 = 114$

    $\square = 19$ Simplify.

Check whether the value is 19 or not.

$\square \times 6 = 114$ Original equation

$19 \times 6 = 114$ Replace $\square$ with 19.

$114 = 114$ It means $\square = 19$ is true.

Exercises 19   Solve each equation.

1) $8 \times n = 56$ 2) $3 \times n = 36$

3) $y \times 6 = 42$ 4) $5 \times x = 55$

5) $z \times 5 = 40$ 6) $9 \times n = 72$

7) $3 \times a = 36$ 8) $7 \times c = 56$

9) $3 \times n = 21$ 10) $6 \times n = 30$

11) $y \times 6 + 6 = 60$ 12) $4 \times x - 4 = 24$

Exercises 20   Find the value of each expression.

1) If $n = 6$, find the value of $12 \times n$.   2) If $c = 2$, find the value of $c \times 43 - 24$.

3) If $x = 7$, find the value of $21 - 2 \times x$.   4) If $y \times 3 = 9$, find the value of $y \times 8 + 2$.

5) If $2 \times x = 18$, find the value of $x \times 2 - 5$.   6) If $x \times 5 = 8$, find the value of $5 \times x - 5$.

Exercises 21   Solve each equation.

1) $16 \times n = 56$ **2)** $11 \times n = 77$

_____   _____

3) $y \times 12 = 96$ **4)** $12 \times x = 48$

_____   _____

5) $9 \times \Delta = 144$ **6)** $\Delta \times 18 - 2 = 34$

_____   _____

7) $\Delta \times 6 - 4 = 44$ **8)** $8 + 6 \times \Delta = 86$

_____   _____

9)   **10)**   **11)**

$$\begin{array}{r} 92 \\ \times \boxed{} \\ \hline 736 \end{array}$$ $$\begin{array}{r} 32 \\ \times \boxed{} \\ \hline 128 \end{array}$$ $$\begin{array}{r} \boxed{} \\ \times \ \ 46 \\ \hline 184 \end{array}$$

Exercises 22   Find the value of each expression.

1) If $1 + n = 6$, find the value of $6 \times (n + 1)$.   **2)** If $1 - x = 4$, find the value of $1 \times (1 - x)$.

_____   _____

3) If $3 \times x = 15$, find the value of $15 - 2 \times x$.   **4)** If $y \times 7 = 9$, find the value of $9 + 7 \times y$.

_____   _____

5) If $3 - x = 1$, find the value of $6 \times (3 - x)$.   **6)** If $5 + x = 8$, find the value of $2 \times (x + 5)$.

_____   _____

7) If $x - 5 = 3$, find the value of $x \times (5 - 3)$.   **8)** If $2 \times x = 4$, find the value of $2 \times x - 2$.

_____   _____

3-10. Solve the equation.

$$x \div 5 = 60$$

SOLUTION

The left side of the equation has the same value as the right side.

For solving x, you should multiply each side by 5 and then cancel out 5 on the left side. So x on the left side will remain and then 300 will be on the right side.

* A division can be used a fraction like that easily manipulates it.

$x \div 5 = 60$ or $\frac{x}{5} = 60$ $x \div 5 \Rightarrow \frac{x}{5}$ or $x \div 5 \Rightarrow \frac{1}{5}x$

$\frac{x}{5} = 60$ Original equation

$5 \times \frac{x}{5} = 60 \times 5$ Multiply each side by **5**.

$^1 5 \times \frac{x}{5_1} = 60 \times 5$ Cancel out 5 on the left side.

$x = 300$ Multiply 60 by 5.
So, the solution of $x \div 5 = 60$ is 300.

Check whether x is 300 or not
Once you solved the equation to get $x = 300$, check your solution by substituting the value in the equation.

$x \div 5 = 60$ Original equation
$300 \div 5 = 60$ Replace x with 300.
$60 = 60$ It means $x = 300$ is true.

Exercises 23   Solve each equation.

1) $54 \div n = 9$ 2) $28 \div n = 4$

_____   _____

3) $x \div 6 = 3$ 4) $4 \div n = 2$

_____   _____

5) $56 \div d = 8$ 6) $c \div 2 = 12$

_____   _____

Exercises 24   Solve each equation.

1) $x \div 5 = 9$

2) $7 \div n = 1$

_____

_____

3) $21 \div n = 7$

4) $48 \div n = 4$

_____

_____

5) $25 \div d = 5$

6) $c \div 7 = 9$

_____

_____

7)

$$n\overline{)64}^{\,8}$$

8)

$$5\overline{)75}^{\,d}$$

9)

$$8\overline{)104}^{\,n}$$

10)

$$x\overline{)306}^{\,51}$$

11)

$$y\overline{)84}^{\,6}$$

12)

$$7\overline{)49}^{\,a}$$

Exercises 25   Find the value of each variable.

1) If $x = 2$, find the value of $8 \div x - 3$.

2) If $z = 24$, find the value of $z \div 3 - 2$.

_____

_____

3) If $k = 2$, find the value of $32 \div k + 6$

4) If $n = 9$, find the value of $(56 - 2) \div n$.

_____

_____

5) If $k = 6$, find the value of $6 + 42 \div k$

6) If $p = 5$, find the value of $30 \div 2 - p$.

_____

_____

7) If $w \div 1 = 2$, find the value of $8 \div w$.

8) If $k \div 3 = 3$, find the value of $k \div 3 - 1$.

_____

_____

Exercises 26   Solve the equation.

1) $x \div 4 = 9$

2) $a \div 8 = 8$

3) $105 \div n = 15$

4) $91 \div n = 13$

5) $x \div 5 - 5 = 12$

6) $64 \div n + 5 = 7$

7)
$$n \overline{)84} = 7$$

8)
$$8 \overline{)120} = \square$$

9)
$$12 \overline{)108} = n$$

10)
$$\square \overline{)270} = 27$$

11)
$$\square \overline{)108} = 18$$

12)
$$26 \overline{)390} = \square$$

Exercises 27   Find the value of each variable.

1) If $x = 2$, find the value of $10 \div x - 3$.

2) If $z = 15$, find the value of $z \div 3 - 2$.

3) If $k = 4$, find the value of $4 \div k + 6$.

4) If $n - 2 = 6$, find the value of $16 \div n$.

5) If $k = 26$, find the value of $104 \div k$.

6) If $p = 6$, find the value of $114 \div p - 6$.

7) If $12 \div w = 3$, find the value of $16 - 16 \div w$.

8) If $k \div 2 = 5$, find the value of $k \div 2 + 5$.

Exercises 28   Solve each equation.

1) $81 \div \Delta = 3$

2) $\Delta \div 15 = 3$

3) $64 \div \Delta = 4$

4) $\Delta \div 32 = 5$

5) $\Delta \div 17 = 4$

6) $126 \div \Delta = 6$

7) $108 \div d - 6 = 12$

8) $192 \div d + 2 = 10$

9)
$$8\overline{)}\quad^{58}$$

10)
$$\square\overline{)252}\quad^{42}$$

11)
$$15\overline{)975}\quad^{\square}$$

Exercises 29   Find the value of each variable.

1) If $1 + x = 3$, find the value of $4 \div x - 1$.

2) If $3 - z = 1$, find the value of $(z + 1) \div 3 + 2$.

3) If $k \div 4 = 2$, find the value of $2 + k \div 4$.

4) If $4 \div n = 2$, find the value of $6 - 2 \div n$.

5) If $12 \div k = 4$, find the value of $12 \div 4 + k$.

6) If $p \div 5 = 3$, find the value of $p \div 3 - 1$.

7) If $w \div 4 = 2$, find the value of $(1 + 3) \div w$.

8) If $18 \div k = 3$, find the value of $1 + k \div 3$.

＊ Solving Problem

Exercises 30   Solve the problem using the given information.
Suzan paid $10.35 for three chocolate ice creams. How much money would she need to pay for 12 chocolate ice creams?

SELF-TEST

1. Which of the following expressions represents "the product of 18 and x"?

    A. $18 \div x$ **B.** $18 - x$
    C. $18 + x$ **D.** $18 \times x$

2. Which of the following expressions represents "the product of N and 1"?

    A. $N + 1$ **B.** $N \div 1$
    C. N **D.** $N - 1$

3. What is the value of x for the equation below?
    $$9 \times x = 153$$

    A. 23   **B.** 35
    C. 32   **D.** 17

4. Marcus has 12 M&Ms that he is dividing equally into 3 piles. Which equation could be used to show how many M&Ms are in each pile?

    A. $3 = 12 \times x$ **B.** $y = 2x + 12$
    C. $12 = 3x$ **D.** $12 = 3 \times x + 3$

5. The bakery is baking loaves of banana bread. They have 40 bunches of banana. If there are 8 bananas to a bunch, what is the equation showing how many bananas there are?

    A. $40 = x + 8$ **B.** $40 = 8x$
    C. $40 = x \div 8$ **D.** $8 = 40 \times x$

6. A miser is counting his coins and organizing them into bags. He has organized his

coins so that there are 3 coins in a bag. If the miser has 24 coins in total, what equation could be used to show how many bags he has?

A. $3 = x \div 24$

B. $24 = x \div 3$

C. $24 \times 3 = x$

D. $24 = 3x$

7. At a fundraiser, 11 people have each donated the same amount of money. If the money they donated is $187, how did you set up the equation that shows how much money each person donated?

A. $11 = x \times 187$

B. $187 = x \div 11$

C. $11x = 187$

D. $187 \times 11 = x$

8. A table is set out for a buffet. On the table are a few plates of crab. Each plate has exactly 5 pieces of crab. There are 56 pieces of crab in total. What is the equation showing how many plates of crab are set up?

A. $5 \times 56 = x$

B. $56 = x \div 5$

C. $56x = 5$

D. $56 = 5x$

9. Which of the following expressions represents "18 divide by x"?

A. $18 \div x$

B. $18 - x$

C. $18 + x$

D. $18 \times x$

10. Which of the following expressions represents "y divide by 3"?

A. $3 \div y$

B. $y \div 3$

C. $3 + y$

D. $y \times 3$

11. What is the value of x for the equation below?
$$72 \div x = 9$$

A. 7

B. 8

C. 9

D. 10

12. What is the value of x for the equation below?
$$x \div 16 = 6$$

A. 23   B. 96
C. 32   D. 19

* For Exercises **13-14**, Bob has 85 toy cars that he keeps in several boxes. He likes to keep 17 toy cars in each box.

13. Which of the following equations could be used to find how many boxes he used?

A. $85 \div x = 17$ B. $85 \times 17 = x$
C. $85 \times x = 17$ D. $17 \div 85 = x$

14. How many boxes does he use?

A. 4   B. 5
C. 6   D. 7

15. The school is planning a field trip. If there are 200 students and 50 seats in a school bus, which equation could be used to determine how many buses are needed to take every student?

A. $50 \div x = 200$ B. $200 \times x = 50$
C. $200 \div x = 50$ D. $50 \div 200 = x$

16. How many buses will be needed to take every student?

A. 1   B. 2
C. 3   D. 4

17. Given an equation with a quotient of 2 and a dividend of 18, what is the value of the divisor?

A. 36   B. 18
C. 9   D. 2

18. If $k = 15$, what is the value of $27 - (k + 6)$?

A. 6   B. 8
C. 10   D. 12

19. If $z = 2$, what is the value of $(3 - z) + 6$?

A. 5   **B.** 6
C. 7   **D.** 8

20. Given the quotient of 18 and x increased by 8, what is the value if $x = 9$?

    A. 9   **B.** 10
    C. 11   **D.** 12

21. The product of 7 and K decreased by 2, what is the value if $K = 5$?

    A. 10   **B.** 0
    C. 33   **D.** 20

22. Given the sum of 19 and N increased by 5, what is the value if $x = 2$?

    A. 26   **B.** 23
    C. 13   **D.** 41

23. Given the difference between x and 3 divided by 2, what is the value if $x = 15$?

    A. 8   **B.** 30
    C. 9   **D.** 6

24. If $z = 7$, what is the value of $2z + 6$?

    A. 14   **B.** 20
    C. 15   **D.** 8

25. If $x = 3$, what is the value of $(9 - 2x) \times 6$?

    A. 24   **B.** 18
    C. 16   **D.** 20

CHAPTER 4
Measurement and Geometry

In this chapter, you will identify the various angles and find perimeters and areas of square, rectangle, and triangle. You will classify the triangles and parallelograms and find the volume of solids.

1. Identifying Angles

4-1. Measurement

* Definition.

1) Acute angle: An angle whose measure is less than 90°.
2) Right angle: An angle that is 90°
3) Obtuse angle: An angle that is between 90° and 180°.
4) Perpendicular lines: Intersecting lines that form a right angle.
5) Parallel lines: Lines that will never intersect.
6) Intersecting lines: Lines that intersect but do not form a right angle

Exercises 1   Identify.

A. Acute angle   B. Right angle   C. Perpendicular lines
D. Parallel lines   E. Obtuse angle   F. Intersecting lines

1)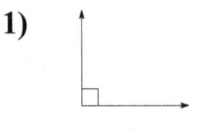

2)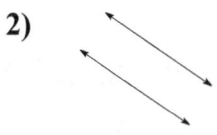

3)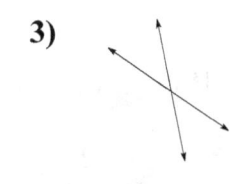

4)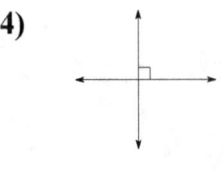

5)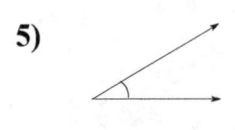

6)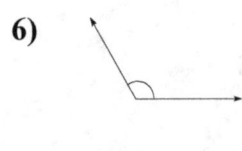

1. Which of the following classifies lines that never intersect?

    A. Obtuse angle   B. Parallel lines
    C. Intersecting lines   D. Perpendicular lines

2. Which of the following figures are intersecting lines?

    A. 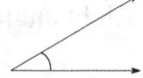   B.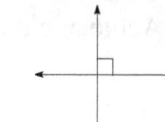

    C.    D.

3. Which of the following classifies an angle that is less than 90°?

    A. Obtuse angle   B. Scalene triangle
    C. Acute angle   D. Equilateral triangle

4. Which of the following figures is an obtuse angle?

    A.    B.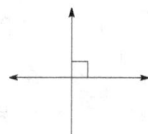

    C. 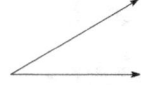   D.

5. Which of the following figures are perpendicular lines?

    A.    B.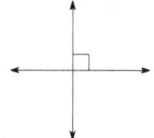

    C.    D.

6. Which of the following classifies an angle that is greater than 90° and less than 180°?

      A. Obtuse angle   **B.** Scalene triangle
      C. Acute angle   **D.** Equilateral triangle

7. Which of the following classifies an angle that is 90°?

      A. Obtuse angle   **B.** Scalene triangle
      C. Acute angle   **D.** Right angle

8. Which of the following is the correct definition of an obtuse angle?

      A. An angle that is 90°
      B. An angle with a measure between 90° and 180°.
      C. Two lines form a right angle.
      D. Two lines that will never intersect.

2. Perimeters and Areas

4–2. Perimeter (P) and area of a polygon

      * To find the perimeter of a polygon, add the length of each side.

Rectangle	Square	Triangle
Perimeter (P) = 2(length + width) Area = length x width $A = \ell w$	Perimeter (P) = 4s Area = side x side $A = s^2$	Perimeter (P) = c + b + h Area = $\frac{1}{2}$ (base x height) $A = \frac{1}{2}bh$

Exercises 2   Find the perimeter of each figure.

1)

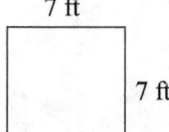

7 ft

7 ft

2)

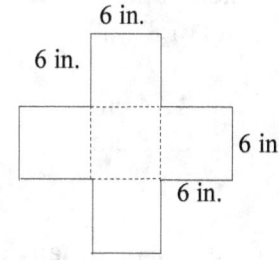

12 m

7 m

3)

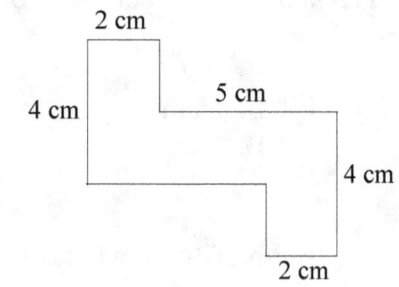

2 cm

4 cm

5 cm

4 cm

2 cm

4)

6 in.

6 in.

6 in.

6 in.

5)

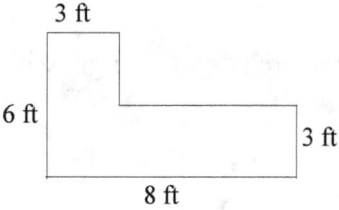

3 ft

6 ft

3 ft

8 ft

6)

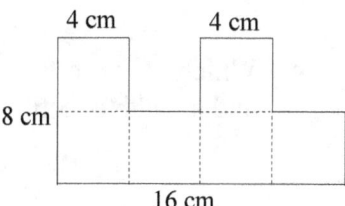

4 cm   4 cm

8 cm

16 cm

7)

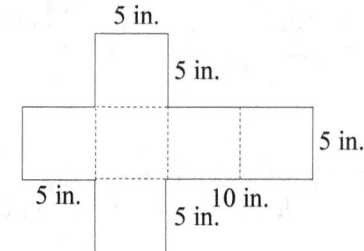

5 in.

5 in.

5 in.

5 in.   10 in.

5 in.

8)

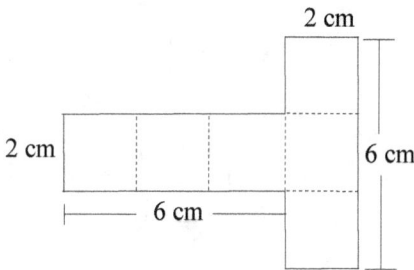

2 cm

2 cm

6 cm

6 cm

SELF-TEST

1. What is the perimeter for the figure below?

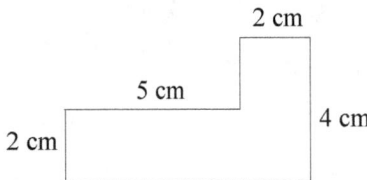

A. 14 cm	**B.** 16 cm
C. 18 cm	**D.** 20 cm

2. Which of the following is the length of a rectangle with a perimeter of 34 cm and a width of 8 cm?

    A. 4 cm   **B.** 5 cm
    C. 8 cm   **D.** 9 cm

3. If the area of a triangle is 28 m^2 with a base length of 14 m, what is the height of the triangle?

    A. 4 m   **B.** 8 m
    C. 12 m   **D.** 16 m

4. Which of the following is the perimeter and area of a rectangle with a length of 10 ft and a width of 5 ft?

    A. P = 50 ft, A = 30 ft^2 **B.** P = 30 ft, A = 50 ft^2
    C. P = 15 ft, A = 35 ft^2 **D.** P = 30 ft, A = 100 ft^2

5. Find the perimeter of the rectangle in the diagram below.

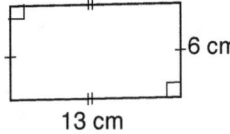

A. 19 cm	**B.** 29 cm
C. 38 cm	**D.** 78 cm

6. Which of the following are the perimeter and area of a triangle given that the side lengths are 2 cm, 3 cm, and 4 cm, with a base length of 4 cm and a height of 2 cm long?

    A. P = 11 cm, A = 12 cm^2 **B.** P = 9 cm, A = 8 cm^2
    C. P = 9 cm, A = 4 cm^2 **D.** P = 12 cm, A = 4 cm^2

7. If the base of a triangle is 12 ft long and the height is 6 ft long, which of the following is the area of the triangle?

      A. 18 ft^2 B. 28 ft^2

      C. 36 ft^2 D. 72 ft^2

8. If a triangle has side lengths of 9 ft., 10 ft., and 5 ft., which of the following is the perimeter of the triangle?

      A. 12 ft   B. 27 ft

      C. 24 ft   D. 29 ft

9. Which of the following is the area of a rectangle given that its length is 12 cm and its width is 7 cm?

      A. 19 cm^2 B. 38 cm^2

      C. 84 cm^2 D. 7056 cm^2

10. If area of two triangles is 12 ft^2, what is the perimeter for the parallelogram below?

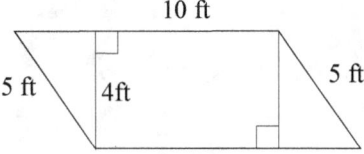

      A. 33 ft   B. 34 ft

      C. 35 ft   D. 36 ft

11. What is the perimeter for the figure below?

      A. 12 inches   B. 36 inches

      C. 72 inches   D. 24 inches

12. Which of the following is the area of a triangle given that the base is 12 inches and the height is 6 inches?

      A. 18 in^2 B. 6 in^2

      C. 36 in^2 D. 72 in^2

3. Classifying Triangles

4-3. What is congruency?

Congruency: If figures are congruent, that means they have the same shape, size, and angle.

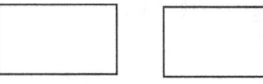

congruent

4-4. Classify the triangle.

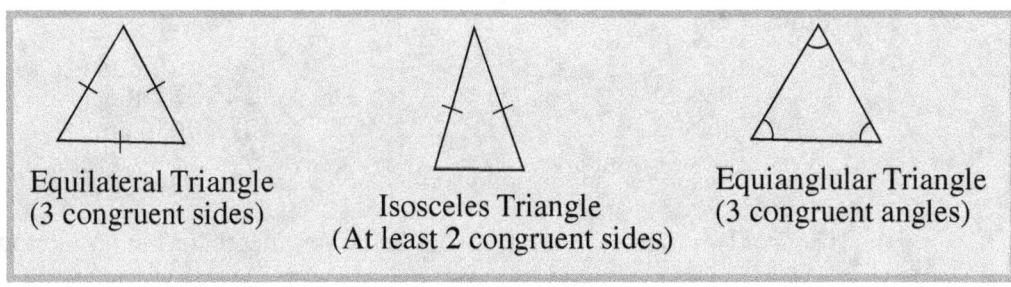

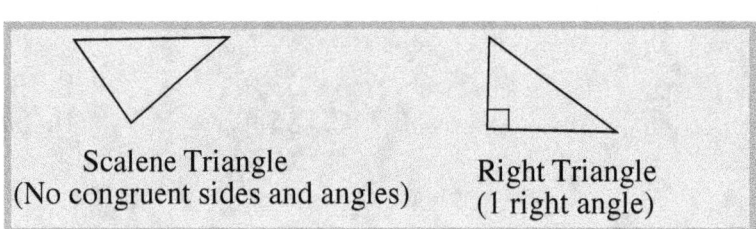

SELF-TEST

1. Which of the following shapes are congruent?

A.   B.

C.   D.

2. Which of the following is a scalene triangle?

A.
6 in.   6 in.
8 in.

B.
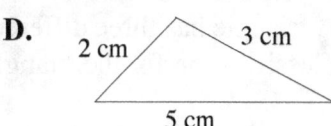

C.
4 ft   4 ft
4 ft

D.
2 cm   3 cm
5 cm

3. Which of the following is an isosceles triangle?

A. $64°, 56°, 60°$ B. $60°, 60°, 60°$
C. $48°, 66°, 66°$ D. $42°, 64°, 72°$

4. A triangle has side lengths of 12 ft, 10 ft, and 8 ft. Which of the following is the correct classification for the triangle?

A. Equilateral triangle   B. Scalene triangle
C. Right triangle   D. Isosceles triangle

5. Which of the following shapes is an isosceles trapezoid?

A.   B.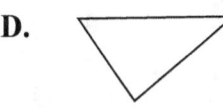

C.   D.

6. All the angles of a triangle are congruent. Which of the following is the correct classification for the triangle?

A. Scalene triangle   B. Right triangle
C. Equiangular triangle   D. Isosceles triangle

7. If two sides of a triangle equal 17 cm, which of the following is the correct classification for the triangle?

    A. Right triangle   **B.** Equilateral triangle
    C. Isosceles triangle   **D.** Scalene triangle

8. A triangle has three different angles. Which of the following is the correct classification for the triangle?

    A. Equilateral triangle   **B.** Isosceles triangle
    C. Scalene triangle   **D.** Right triangle

9. A triangle has one side that is different from the rest. Which of the following is the correct classification for the triangle?

    A. Scalene triangle   **B.** Right triangle
    C. Equilateral triangle   **D.** Isosceles triangle

10. Which of the following shapes is a scalene triangle?

    A. 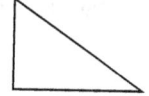   **B.**

    C. 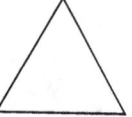   **D.**

11. Which of the following is the correct definition of an acute triangle?

    A. All three angles are less then 90°.
    B. Two angles of the triangle are congruent.
    C. Two sides of the triangle are congruent.
    D. The triangle has three different sides.

12. Which of the following classifies a triangle with three congruent sides?

    A. Scalene triangle   **B.** Acute triangle
    C. Equilateral triangle   **D.** Isosceles triangle

13. Which of the following classifies a triangle with three different sides?

    A. Equilateral triangle   **B.** Scalene triangle
    C. Isosceles triangle   **D.** Acute triangle

14. Which of the following shapes is an equiangular triangle?

    A. 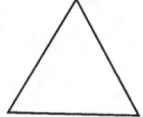   **B.**

    C.    **D.**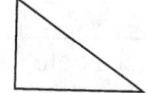

15. Which of the following classifies a triangle with at least 2 congruent sides?

    A. Equilateral triangle   **B.** Isosceles triangle
    C. Equiangular triangle   **D.** Acute triangle

16. Which of the following classifies the triangle shown below?

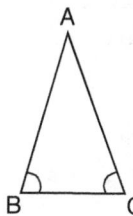

    A. Isosceles triangle   **B.** Scalene triangle
    C. Equiangular triangle   **D.** Equilateral triangle

17. What is the area for a triangle below?

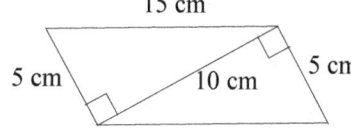

    A. 50 cm^2 **B.** 75 cm^2
    C. 25 cm^2 **D.** 40 cm

4. Classifying Quadrilaterals
4–5. Classifying quadrilaterals

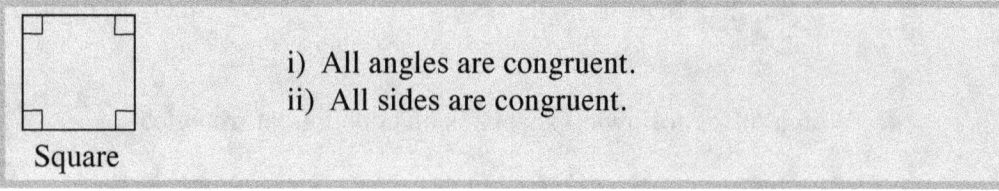

i) All angles are congruent.
ii) All sides are congruent.

Square

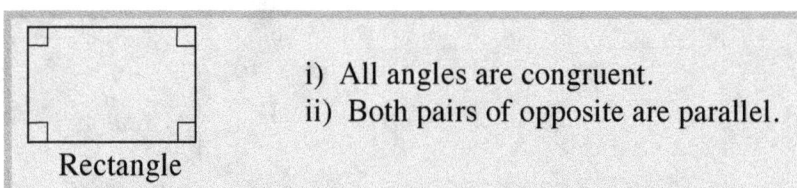

i) All angles are congruent.
ii) Both pairs of opposite are parallel.

Rectangle

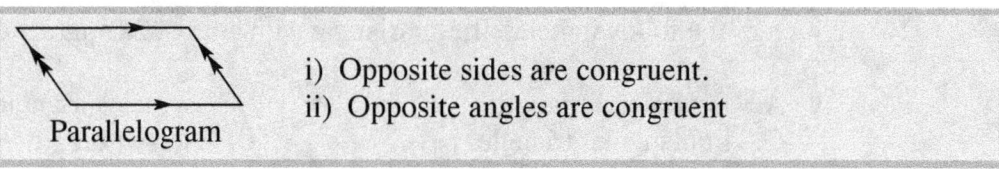

i) Opposite sides are congruent.
ii) Opposite angles are congruent

Parallelogram

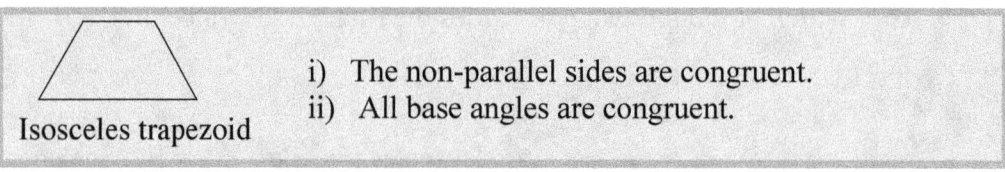

i)   The non-parallel sides are congruent.
ii) All base angles are congruent.

Isosceles trapezoid

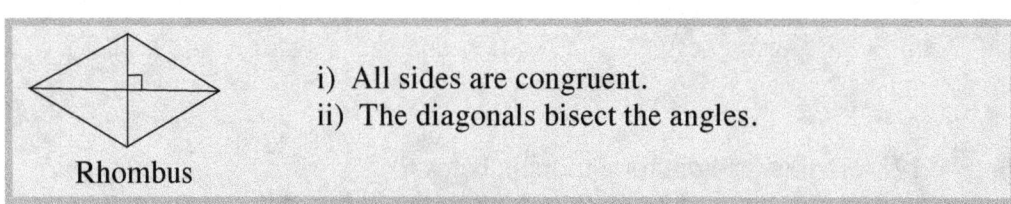

i) All sides are congruent.
ii) The diagonals bisect the angles.

Rhombus

1. Which of the following statements is NOT a property of a rectangle?

    A. Diagonals that bisect each other always form four isosceles triangles.
    B. A rectangle has two pairs of opposite sides.
    C. The sum of the angles formed by the diagonals is 360°.
    D. A rectangle is always a parallelogram.

2. Which of the following is always a true statement?

    A. All angles of a rhombus are congruent.
    B. All angles of an isosceles trapezoid are congruent.
    C. All sides of a rectangle are congruent.
    D. A square is a parallelogram.

3. Which of the following is not a parallelogram?

    A. 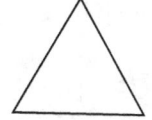   **B.**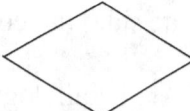

    C. 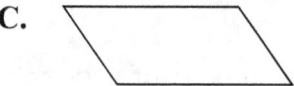   **D.**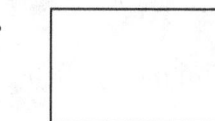

4. Which of the following is always a true statement?

    A. Opposite angles of a rectangle are congruent.
    B. The sum of the inside angles of a rectangle is $360°$.
    C. All angles of a rectangle are congruent.
    D. All of the above.

5. Which of the following is a true statement?

    A. A rhombus is a rectangle.   **B.** A rhombus is a square.
    C. A rhombus is a parallelogram.   **D.** All of the above.

6. Which of the following shapes do not have angles that are all congruent?

    A. Rectangle   **B.** Square
    C. Rhombus   **D.** Parallelogram

7. If a shape has opposite angles, they are congruent. Which of the following shapes do not have this property?

    A. Rectangle   **B.** Isosceles trapezoid
    C. Rhombus   **D.** Parallelogram

8. Which of the following shapes is a rhombus?

A.

B.

C.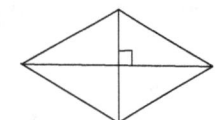

D.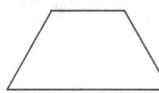

9. Opposite sides are congruent. Which of the following shapes are true?

    A. Parallelogram   **B.** Isosceles trapezoid
    C. Rhombus   **D.** All of the above

10. Which of the following statements is NOT a property of a square?

    A. Diagonals that bisect each other always form four isosceles triangles.
    B. The sum of interior angles is 360°
    C. All of the sides of a square are congruent.
    D. All of the angles of a square are congruent.

11. Which of the following is not a true statement?

    A. The diagonals of a rhombus are always congruent.
    B. All angles of a rhombus are congruent.
    C. A rhombus is always a parallelogram.
    D. A rhombus is always a quadrilateral.

12. Which of the following is always a true statement?

    A. All sides of a rectangle are congruent.
    B. A square is a parallelogram.
    C. All angles of a rhombus are congruent.
    D. All of the above.

13. Which of the following statements is NOT a property of a parallelogram?

    A. A parallelogram has congruent opposite angles.
    B. A parallelogram has congruent opposite sides.
    C. The sum of the angles formed by the diagonals is 360°.
    D. A parallelogram is always a rectangle.

14. Which of the following is a true statement?

    A. A square is always a parallelogram.
    B. A square is always a quadrilateral.
    C. All angles of a square are always congruent.
    D. All of the above.

15. Which of the following statements is NOT a property of a rhombus?

    A. The opposite angles of a rhombus are congruent.
    B. A rhombus has two pairs of opposite sides.
    C. The sum of the interior angles is $360°$.
    D. A rhombus always has congruent sides

16. Which of the following quadrilaterals do not have perpendicular diagonals?

    A. Kite   **B.** Square
    C. Rhombus   **D.** All of the above.

17. Which of the following quadrilateral has four congruent sides?

    A. Rectangle   **B.** Right triangle
    C. Rhombus   **D.** Isosceles trapezoid

18. Which of the following statements is NOT a property of an isosceles trapezoid?

    A. An isosceles trapezoid has congruent opposite angles.
    B. An isosceles trapezoid has congruent base angles.
    C. The sum of the angles formed by the diagonals is $360°$.
    D. The non-parallel sides are always congruent.

5. Finding Volume of Prisms

4-6. Volume of a prism

The volume of the rectangular prism is the product of its length, width, and height.

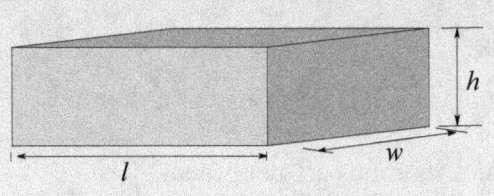

$V = l \times w \times h$, where l is the length, w is the width, and h is the height of the rectagular prism.

4-7. Find the volume of the box.

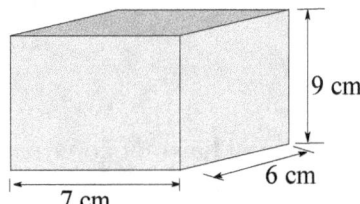

SOLUTION

Use the formula of the volume of the box. $V = l \times w \times h$, where l is the length, w is the width, and h is the height of the box. $V = 7 \text{ cm} \times 6 \text{ cm} \times 9 \text{ cm} = 378 \text{ cm}^3$ So the volume is 378 cm^3. Remember that when writing the volume, put it in cubic units or units^3.

Exercises 3   Name each figure.

A. Pyramid   B. Sphere   C. Rectangular prism   D. Cone

1)   2)   3)   4)

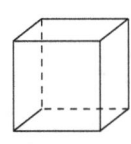   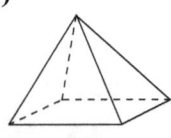      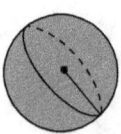

_____   _____   _____   _____

Exercises 4   Solve each problem using the given information.

1)   Find the volume of a rectangular prism that has a height of 32 cm, a length of 18 cm, and a width of 6 cm.

2)   Find the volume of a cube that has a side length of 7 in.

3)   Find the volume of a rectangular prism that has a length of 15 mm, a width of 2 mm, and a height of 24 mm.

4)   Find the height of a rectangular prism that has a volume of 135 cm^3, a length of 9 cm, and a width of 3 cm.

5)   Find the length of a rectangular prism that has a volume of 264 ft^3, a height of 6 ft, and a width of 4 ft.

6)   Find the volume of a rectangular prism that has a side length of 4 ft, a width of 9 ft, and a height of 2 ft.

7)   Find the width of a rectangular prism that has a volume of 192 $yards^3$, a length of 6 yards, and a height of 8 yards.

8)   Find the length of a rectangular prism that has a volume of 120 in^3, a height of 5 in, and a width of 2 in.

9)   If the width of a rectangular prism is 30 ft, what is the volume, given that it has a height of 4 ft and a length of 7 ft?

1. Find the volume of the box below.

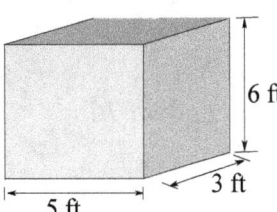

A. 18 ft^2 **B.** 13 ft^2
C. 18 ft^3 **D.** 90 ft^3

2. What is the volume of the rectangular prism below?

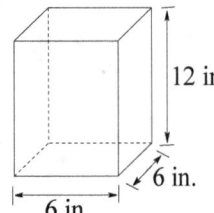

A. 288 in^3 **B.** 72 in^2
C. 72 in^3 **D.** 432 in^3

3. What is the volume of the rectangular prism below?

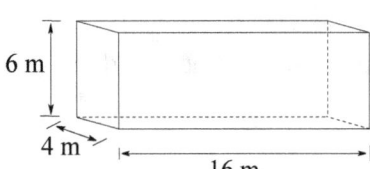

A. 384 m^3 **B.** 240 m^3
C. 480 m^3 **D.** 26 m

4. Find the volume of the rectangular prism. Round your answer to the nearest tenth if necessary.

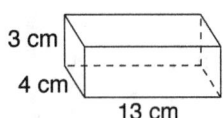

A. 78 cm^3 **B.** 156 cm^3
C. 312 cm^3 **D.** 624 cm^3

5. Which of the following is the volume of a box with a width of 2 cm, a length of 5 cm and a height of 3 cm long?

A. 150 cm^3 **B.** 30 cm^2
C. 30 cm^3 **D.** 15 cm^2

6. If the length of a box is 8 ft long and the height is 6 ft long, which of the following is the volume of the rectangular box?

    A. 120 ft^2 **B.** 120 ft^3
    C. 240 ft^3 **D.** 240 ft^2

7. If a box has a length of 3 in., a width of 2 in., and a height of 5 in., which of the following is the volume?

    A. 12 ft   **B.** 27 ft
    C. 30 ft   **D.** 32 ft

8. Which of the following is the height of a rectangle with a volume of 112 cm^3, a length of 7 cm, and a height is 4 cm?

    A. 2 cm   **B.** 4 cm
    C. 6 cm   **D.** 8 cm

9. Which of the following solids is a pyramid?

    A. 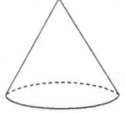   **B.**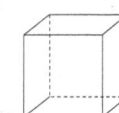

    C. 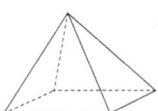   **D.**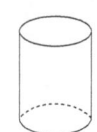

10. What is the volume of a rectangular prism with a height of 32 cm, a length of 18 cm, and a width of 6 cm?

    A. 6912 cm^3 **B.** 1536 cm^3
    C. 1728 cm^3 **D.** 3456 cm^3

11. If the volume of the rectangular prism is 864 ft^3, what is the length of the rectangular prism?

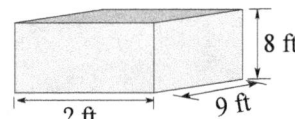

A. 12 ft   **B.** 10 ft
C. 8 ft   **D.** 14 ft

12. What is the volume of the rectangular prism below?

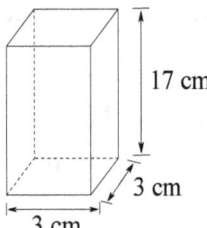

17 cm

3 cm

3 cm

A. 204 cm^3 **B.** 204 cm
C. 153 cm^3 **D.** 153 cm

13. Which of the following solids is a cone?

A. 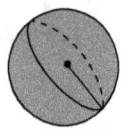   **B.**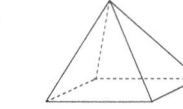

C. 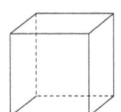   **D.**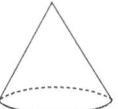

14. If the volume of a cube is 216 ft^3, what is the height of the cube?

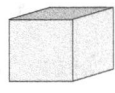

V = 216 ft^3

A. 3 ft   **B.** 4 ft
C. 5 ft   **D.** 6 ft

6. Understanding Coordinate Grids

4–8. Plot A(4, 3) on a coordinate grid.

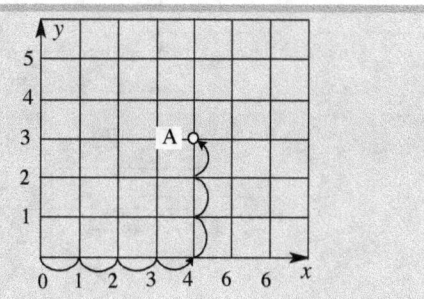

First, look at A(4, 3).
   (*x*-coordinate, *y*-coordinate)

i)   Start at point 0.
ii) Count 4 units to the left along the *x*-axis.
iii) Then count 3 units up the *y*-axis.

Exercises 5   Plot the points in a coordinate plane and connect them together.

1) A(10, 10), B(4, 2), C(10, 4)   **2)** P(1, 5), Q(6, 5), R(1, 2), S(6, 2)

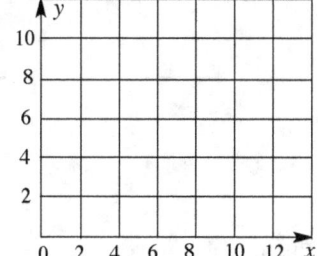   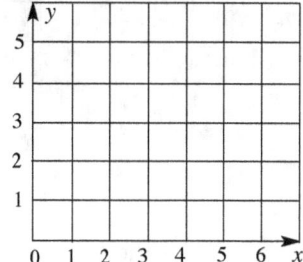

3) D(0, 4), E(4, 4), F(0, 0)   **4)** T(6, 0), U(0, 4), V(6, 10)

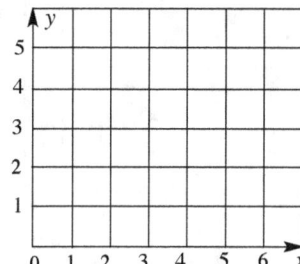   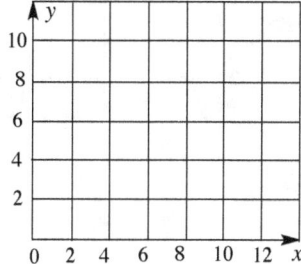

5) G(1, 5), H(1, 2), I(6, 2)   **6)** J(3, 10), K(15, 25), L(18, 5)

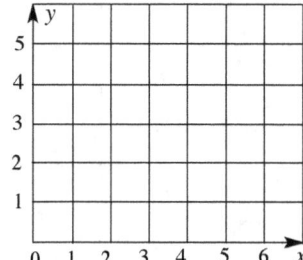   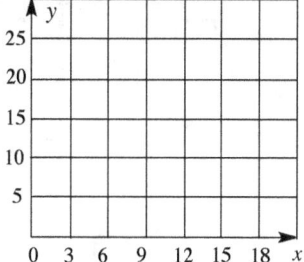

Exercises 6   Find the coordinates that represent M and N.

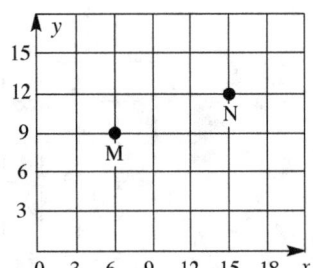

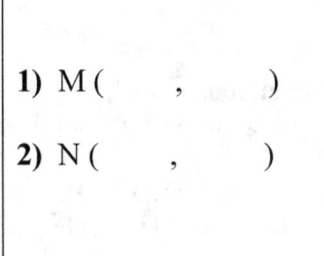

1) M (,   )

2) N (,   )

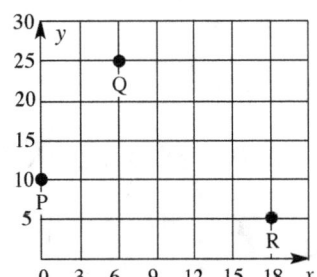

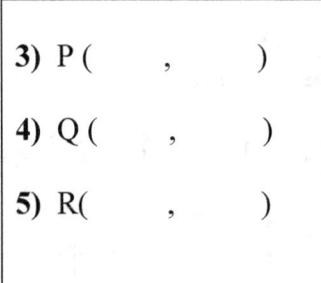

3) P (,   )

4) Q (,   )

5) R(,   )

Exercises 7   Find the length of each line segment.

1)

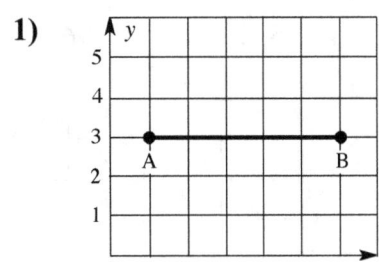

2)

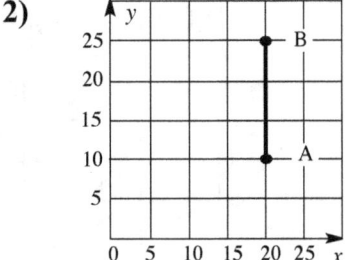

3)

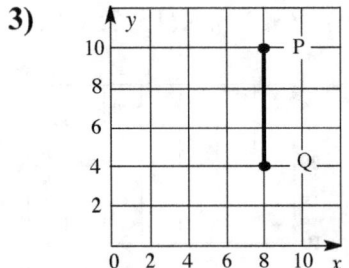

4)

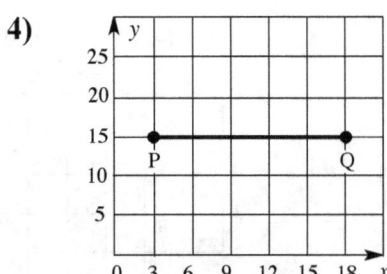

1. What are the coordinates that represent P?

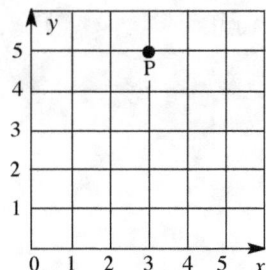

A. P(5, 3)   **B.** P(4, 3)
C. P(3, 5)   **D.** P(3, 3)

2. What are the coordinates that represent K and L?

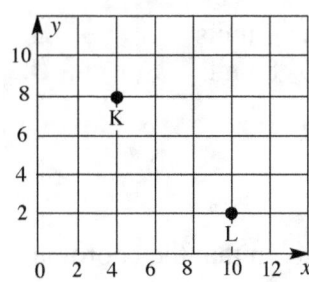

A. K(4, 8), L(10, 2)   **B.** K(4, 8), L(2, 10)
C. K(8, 4), L(2, 10)   **D.** K(8, 4), L(10, 2)

3. What are the coordinates that represent Z?

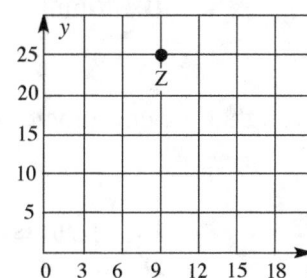

A. Z(3, 5)   **B.** Z(25, 9)
C. Z(9, 25)   **D.** Z(5, 3)

4. Which point is located at the coordinates (4, 3)?

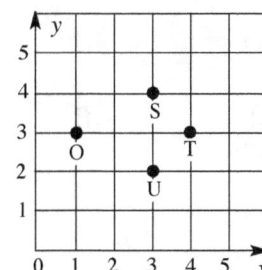

A. U   **B.** S
C. O   **D.** T

5. What is the length of the line segment?

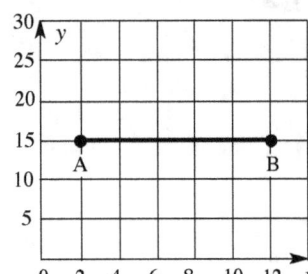

A. 4 units   B. 6 units
C. 8 units   D. 10 units

6. What is the length of the line segment?

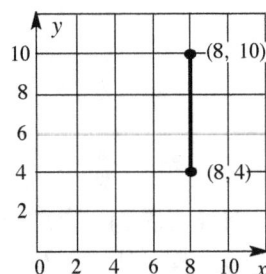

A. 4 units   B. 6 units
C. 8 units   D. 10 units

7. Which of the following is the length of AB given that the coordinates of the endpoints are A(2, 6) and B(9, 6)?

A. 4 units   B. 5 units
C. 6 units   D. 7 units

8. Which of the following is the length of ST given that the coordinates of the endpoints are S(5, 3) and T(5, 14)?

A. 11 units   B. 10 units
C. 9 units   D. 8 units

9. Which of the following is the length of BC?

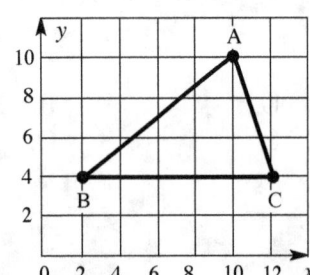

A. 8 units   B. 9 units
C. 10 units   D. 11 units

10. What is the length of AB?

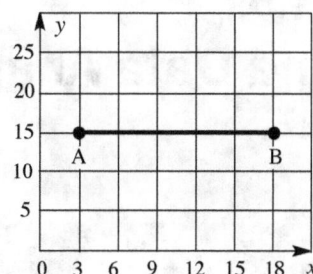

A. 13 units   B. 14 units
C. 15 units   D. 16 units

11. Which of the following is the length for QR?

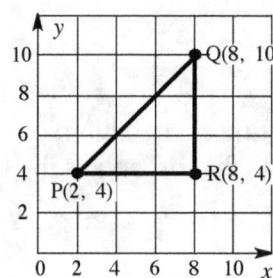

A. 3 units   B. 4 units
C. 5 units   D. 6 units

12. Using the diagram from Question **11**, what is the area of the triangle?

A. 36 units^2 B. 54 units^2
C. 18 units^2 D. 72 units^2

13. Which of the following is the length of the line segment?

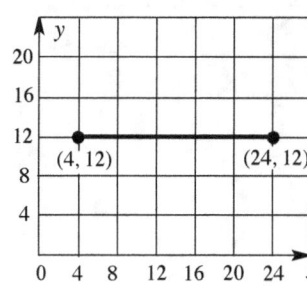

A. 18 units   B. 20 units
C. 22 units   D. 24 units

CHAPTER 5
Probability, Statistics, and Data Analysis

In this chapter, you will learn about frequency and frequency tables and identify the mean, median, and mode of data sets.

1. Frequency and Frequency Tables

5–1. Know the concepts of frequency and frequency tables.

> Frequency and Frequency Tables
> Frequency: The frequency is the number of times a data item occurs.
> Frequency Table: A frequency table is a table that organizes the frequency of the data items.

5–2. A graph that shows 18 students who participated in the survey for their favorite lunch menu.

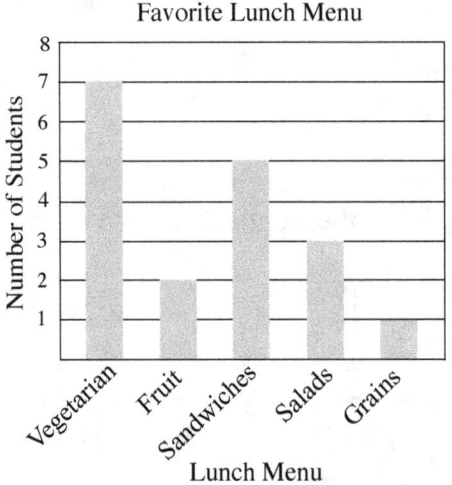

Favorite Lunch Menu

a) Make a frequency table using the data shown above.
b) What are the most and least favorite menu items for lunch?

SOLUTION

a) Make a table with three columns representing the lunch menu, tally, and frequency on the row. First, fill in the items of menu from left to right that shows the graph and then tally the number of items. Count the tally marks and fill in the frequency section.

Lunch Menu	Tally	Frequency
Vegeterian	⻌ ⎟⎟	7
Fruit	⎟⎟	2
Sandwiches	⻌	5
Salads	⎟⎟⎟	3
Grains	⎟⎟⎟	3

b) Based on the table and graph, vegetarian is the most favorite menu item and grains are the least favorite menu item for lunch.

Exercises 1   Use the information for Exercises **1-4**. Lisa asked her classmates their favorite kinds of pets. Using the results, she made a bar graph.

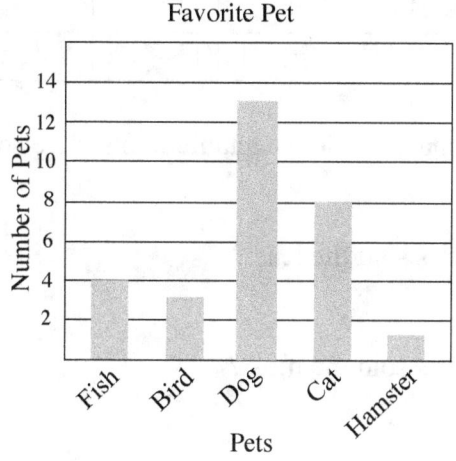

Favorite Pet

1) The table below is missing the tally and frequency section. Complete the table.

Favorite Pet	Tally	Frequency
Fish		
Bird		
Dog		
Cat		
Hamster		

2) Which pet was voted the most favorite? Explain.

3) What is the difference of the votes between dogs and cats?

4) How many people participated in the survey?

Exercises 2   Use the information for Exercises **1-3**. The table shows the results of the number of lunch menu items sold at school.

Menu Sections	Tally	Frequency
Hamburgers	ⵌ |||	8
Sandwiches	ⵌ	5
Hot Dogs	ⵌ ||||	9
Vegetables			1
Fruits	|||	3

1) How many sandwich and vegetable items were sold?

2) Which item was sold the least?

3) Which item was sold the most?

Exercises 3   For Exercises **1-5**, the information below shows the number of hours that Sydney spent doing her homework for science and math.

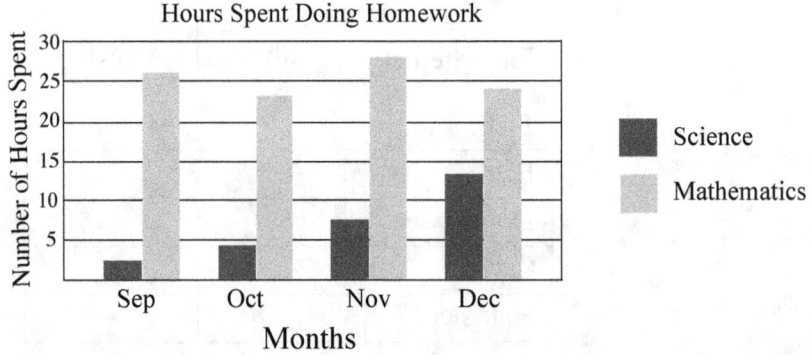

1) Which month had the greatest amount of time spent doing science homework?

2) If Sydney goes to school 25 days per month, how many hours per day does she spend on her math homework in September?

3) Which month did Sydney spent the least amount of time on her science homework?

4) Which month did Sydney spent the least amount of time on her math homework?

5) Describe your thoughts based on the double-bar graph.

2. Mean, Median, and Modes

5–3. Mean, Median, and Mode

Median: The value located in the middle of the data set put in numerical order.
Mode: The value that occurs most often in the data set.

5–4. Find the median and mode of the following data set:
6, 9, 7, 6, 10, 12, 7, 5, 7

SOLUTION

a) Median
    i) Reorder the numbers from least to greatest.
        5, 6, 6, 7, 7, 7, 9, 10, 12
    ii) Find the number in the middle.

$$\begin{matrix} 1 & 2 & 3 & 4 & 5 & 4 & 3 & 2 & 1 \\ 5, & 6, & 6, & 7, & 7, & 7, & 9, & 10, & 12 \end{matrix}$$

↑
Median or middle number

So, the median of the data set is 7.

b) Mode
    i) Reorder the numbers from least to greatest.
        5, 6, 6, 7, 7, 7, 9, 10, 12
    ii) Find the number that occurs most often.
        5, 6, 6, 7, 7, 7, 9, 10, 12

↑
Mode

So, the mode of the data set is 7.
* Some data sets may not have a mode.

Exercises 4   Find the mode and the median in each data set.

1) 9, 15, 8, 13, 11, 16, 9, 12

2) 18, 20, 18, 21, 22, 24, 18, 23

3) 10, 25, 20, 20, 15, 10, 20

4) 65, 69, 62, 67, 66, 70, 61

5) 14, 7, 9, 8, 10, 13, 16, 15

6) 23, 20, 21, 24, 18, 20, 20, 22, 20

Exercises 5   Find the mean and the median in each data set.

1) 0.45, 0.55, 0.35, 0.50

2) 7.3, 4.5, 5.6, 6.8

3) 25, 32, 53, 29, 35, 26, 37, 19

4) 8, 10, 7, 9, 7, 13, 11, 10

5) 5, 3, 7, 3, 6, 5, 7, 5

6) 15, 20, 15, 25, 20, 15, 15, 25

7) 126, 163, 145, 198, 178, 153

8) 1, 4, 2, 7, 3, 6, 4, 2, 4

SELF-TEST

1. Which of the following numbers is the median?
        12, 9, 17, 10, 11, 15, 16, 15, 13

        A. 23   B. 35
        C. 32   D. 19

* Use the information for Exercises **2-3**. Loi recorded the scores of the test he took at
    school.
            23, 25, 24, 22, 23, 24, 24, 24

2. What is the median of the scores?

        A. 23   B. 35
        C. 32   D. 19

3. What is the mode of the scores?

        A. 23   B. 35
        C. 32   D. 19

4. What is the median of the following scores?
        47, 53, 52, 49, 56, 53, 51, 52

        A. 23   B. 35
        C. 32   D. 19

5. Joan is learning to type on a keyboard. She recorded the number of misspelled words
    within 3 minutes, which were 5, 6, 8, 4, 6, 9, 7, 8, and 8. What is the mode of her
    misspelled words?

        A. 23   B. 35
        C. 32   D. 19

6. What was the median of her misspelled words?

        A. 23   B. 35
        C. 32   D. 19

* For Exercises **7-10**, Jessica prepared 10 different kinds of lunches for 40 people. The table below shows the number of calories of the lunches that each person received.

Number of Calories	Tally	Frequency
198	||	2
202	|||| |	6
246	|||	3
265	|||| ||	7
312	||	2

7. Based on the table above, which of the following graphs shown best describes the table above?

A.

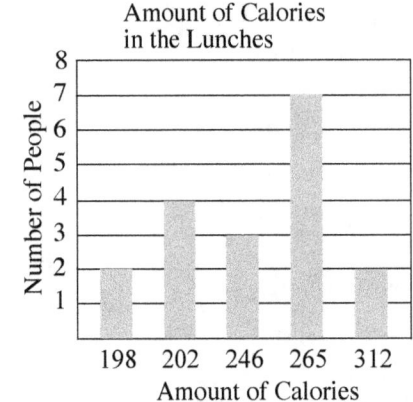

B.

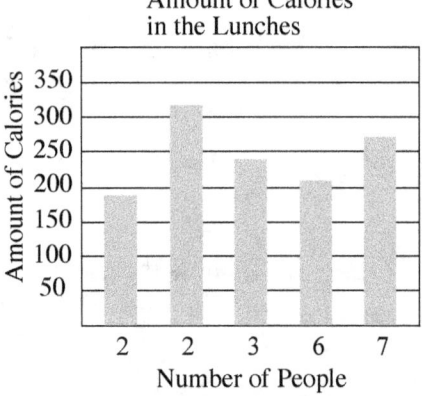

C.

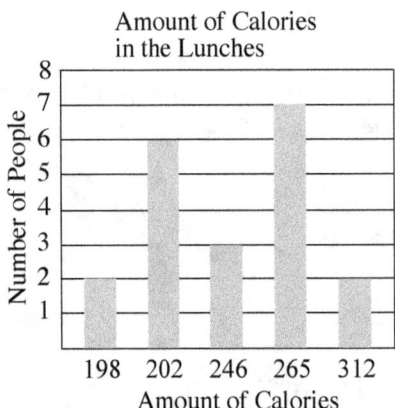

D.

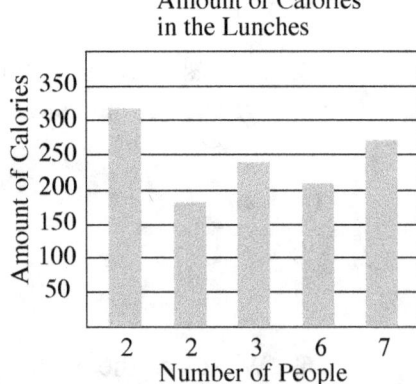

8. How many people participated in Jessica's survey?

    A. 17   **B.** 18

    C. 19   **D.** 20

9. Which lunch was selected by the most people?

      A. 198   **B.** 202
      C. 246   **D.** 265

10. What is the mode of the number of people participating?

      A. 2   **B.** 3
      C. 6   **D.** 7

* For Exercises **11-13**, the graph shows the number of times Jordan visits the public library.

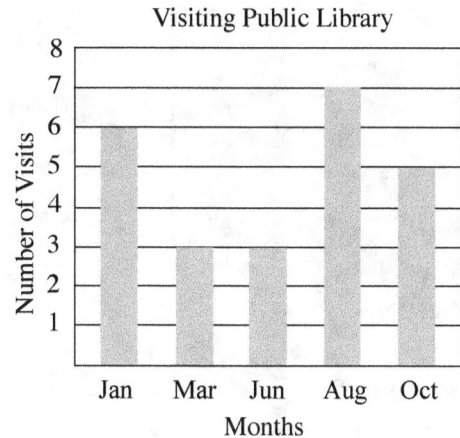

11. Which of the following tables best describes the graph above?

A.

Monthly	Tally	Frequency
Mar	|||	3
Jun	|||	3
Oct	ℍℋ	5
Jan	ℍℋ		6
Aug	ℍℋ ||	7

B.

Monthly	Tally	Frequency
Jan	ℍℋ	6
Mar	|||	3
Jun	|||	3
Aug	ℍℋ	7
Oct	ℍℋ	5

C.

Monthly	Tally	Frequency
Jan	ℍℋ		6
Mar	|||	3
Jun	|||	3
Aug	ℍℋ ||	7
Oct	ℍℋ	5

D.

Monthly	Tally	Frequency
3	|||	Mar
3	|||	Jun
5	ℍℋ	Oct
6	ℍℋ		Jan
7	ℍℋ ||	Aug

12. Which month had the most visits from Jordan?

    A. January   **B.** August

    C. October   **D.** June

13. Which of the following is the mode of the graph?

    A. 3   **B.** 5

    C. 6   **D.** 7

* For Exercises **14-17**, these information and the graph show the average temperatures during winter in Denver and Indianapolis.

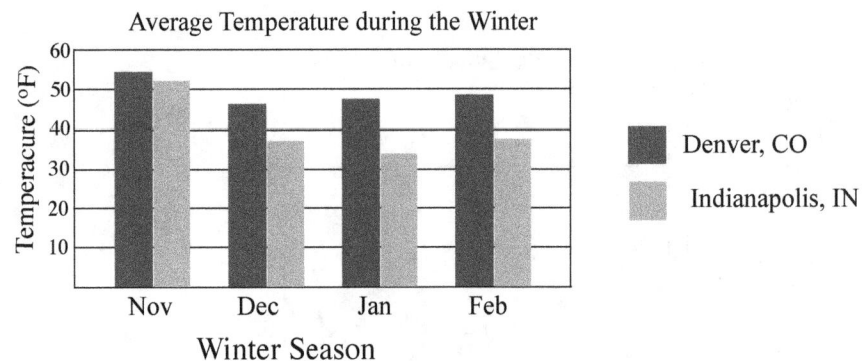

14. What kind of graph is shown above?

    A. Single-bar graph   **B.** Line plot

    C. Line graph   **D.** Double-bar graph

15. Which month had the highest temperature in Denver, Colorado?

    A. January   **B.** August

    C. October   **D.** June

16. Which month had the lowest temperature in Indianapolis, Indiana?

    A. January   **B.** August

    C. October   **D.** June

17. What is the scale of the graph?

    A. January   **B.** August

    C. October   **D.** June

ANSWERS

CHAPTER 1

Exercises 1
1) Thirty-seven
   30 + 7
2) Nine hundred six
   900 + 6
3) Four thousand three
   4,000 + 3
4) Five thousand, forty-eight
   4,000 + 40 + 8
5) Eighty-three thousand, seven hundred forty-six
   80,000 + 3,000 + 700 + 40 + 6
6) Two hundred three million, nine hundred ninety-nine thousand, six hundred twenty
   200,000,000 + 3,000,000 + 900,000 + 90,000 + 9,000 + 600 + 20
7) Five hundred twenty million, seven hundred fifty-two thousand, four hundred sixty-three
   500,000,000 + 20,000,000 + 700,000 + 50,000 + 2,000 + 400 + 60 + 3

Exercises 2
1)	53	2)	207	3)	30,008	4)	535,016,015
5)	96,049	6)	5,400,203	7)	300,811	8)	277,525
9)	81,001,812	10)	100,378				

Exercises 3
1)	-250	2)	-150	3)	50	4)	150

Exercises 4
1)	-16	2)	-8	3)	4	4)	16

Exercises 5
1)	>	2)	>	3)	<	4)	<	5)	>	6)	>

Exercises 6
1) $0 > -20 > -25 > -30 > -45$ 2) $76\,^{\circ}F > 32\,^{\circ}F > -1\,^{\circ}F > -2\,^{\circ}F > -7\,^{\circ}F$
3) $200 > 125 > -5 > -125 > -200$ 4) $25 > 18 > -2 > -19 > -40$

Exercises 7
1) $534,908 > 532,408 > 503,490 > 195,069 > 89,569$
2) $3,276,007 > 27,074,007 > 20,900,007 > 20,746,007$
3) $982,367 > 893,267 > 892,637 > 892,367 > 829,367$

Exercises 8
1)	<	2)	<	3)	<	4)	>	5)	<	6)	>
7)	>	8)	<	9)	>	10)	>	11)	<	12)	<
13)	<	14)	>	15)	<	16)	>				

Answers

Exercises 9
1) 360   2) 8,100   3) 1,050   4) 710,000
5) 84,000   6) 16,100   7) 650,000   8) 260,000
9) 3,800,000   10) 26,100   11) 72,500,000   12) 60,000,000
13) 5,500,000   14) 740,000   15) 400,000   16) 80,000,000

Exercises 10
1) 1,000   2) 1,500   3) C and D, thousand

Exercises 11
1) hundred   2) G   3) thousand

Exercises 12
1) 73,800   2) 74,600   3) G, H, I, ten thousand

Exercises 13
1) 1,500   2) 5,600   3) 1,000,000   4) 24,400
5) 7,000   6) 7,000   7) 6,000,000   8) 40,000
9) 100,000,000   10) 200,000,000

Exercises 14
1) 370,900   2) 6,761,000   3) 47,600   4) 46,400
5) 746,700   6) 6,500

Exercises 15
1) 1,800,000   2) 560,000   3) 46,500,000   4) 24,800,000
5) 700,000   6) 600,000   7) 300,000   8) 37,700,000

Exercises 16
1) 10,000   2) 0   3) 90,000   4) 20,000   5) 130,000   6) 10,000
7) 9,000   8) 0   9) 180,000   10) 50,000

Exercises 17
1) 98,000   2) 6,000   3) 38,000   4) 3,000   5) 96,000   6) 50,000
7) 102,000   8) 7,000   9) 2,000   10) 14,000   11) 18,000

Exercises 18
1) 105,000   2) 70,000   3) 180,000   4) 60,000   5) 110,000   6) 600,000
7) 20,000   8) 10,000   9) 90,000   10) 50,000   11) 170,000   12) 40,000

Exercises 19
1) 406   2) 4,078   3) 711   4) 1,091   5) 1,639   6) 1,208
7) 10,811   8) 6,584   9) 10,003   10) 9,091   11) 4,108   12) 18,599
13) 6,513   14) 6,091   15) 6,486   16) 4,978   17) 6,322   18) 18,896

Exercises 20
1) 377,908   2) 58,811   3) 772,784   4) 1,008,833   5) 13,592   6) 95,685
7) 4,841   8) 95,685   9) 99,017   10) 165,910   11) 10,948   12) 96,465

Exercises 21
1) 4,111,432   2) 1,220,022   3) 2,297,110   4) 110,112
5) 1,187,312   6) 1,222,265   7) 23,112   8) 120,442
9) 511,218

Exercises 22
1) 22   2) 4   3) 241   4) 353   5) 53   6) 5,342
7) 92   8) 278   9) 1,799   10) 4,833

Exercises 23
1) 393   2) 309   3) 889   4) 256   5) 2,449   6) 749
7) 535   8) 97   9) 6,894

Exercises 24
1) 329   2) 1,247   3) 3,839   4) 949   5) 1,589   6) 5,356
7) 290   8) 679   9) 1,741

Exercises 25
1) 1,393   2) 899   3) 928   4) 49,889   5) 18,224   6) 28,069
7) 50,367

Exercises 26
1) 2,949   2) 518,849   3) 6,889   4) 28,987   5) 282,353   6) 48,999
7) 129,792   8) 172,063   9) 8,668   10) 31,963   11) 410,060

Exercises 27
1) 42   2) 15   3) 93   4) 12,921   5) 19   6) 100
7) 2,267   8) 5,676

Exercises 28
1) a) 110,000 & b) 10,000   2) 10,000

Exercises 29
1) 2,809   2) 3,975   3) 4,428   4) 76   5) 17,918   6) $54.00

Exercises 30
1) 100   2) 11 or 12   3) 350   4) 560   5) 3 or 4   6) 9

Exercises 31
1) 1,500   2) 70   3) 1,500   4) 800
5) 200 or 300   6) 300 or 400   7) 2,100   8) 2,400
9) 200 or 300   10) 800 or 900

Exercises 32
1) 1,800   2) 45,000   3) 400   4) 28,000   5) 49,000   6) 70
7) 800 or 900   8) 100   9) 800 or 900

Exercises 33
1) 189   2) 235   3) 175   4) 336   5) 368   6) 288

Exercises 34
1) 497   2) 58   3) 156   4) 168   5) 2,653   6) 3,978
7) 3,033   8) 1,520   9) 2,172   10) 1,820   11) 1,848   12) 1,956

Exercises 35
1) 13   2) 42   3) 76   4) 152   5) 21   6) 618
7) 16   8) 33

Exercises 36

1)	3,696	**2)**	1,525	**3)**	8,466	**4)**	19,257	**5)**	3,704	**6)**	1,808
7)	1,092	**8)**	980	**9)**	15,072	**10)**	3,492				

Exercises 37

1)	468	**2)**	1,175	**3)**	3,510	**4)**	3,654	**5)**	51,428	**6)**	30,408
7)	14,433	**8)**	11,894	**9)**	9,065						

Exercises 38

1)	3,015	**2)**	4,883	**3)**	25,690	**4)**	18,288	**5)**	27,729	**6)**	5,250
7)	9,126	**8)**	49,665	**9)**	21,692						

Exercises 39

1)	9	**2)**	24	**3)**	89	**4)**	137	**5)**	87	**6)**	37
7)	14	**8)**	342	**9)**	826	**10)**	43				

Exercises 40

1)	28	**2)**	37	**3)**	158	**4)**	33

Exercises 41

1)	303	**2)**	153	**3)**	133.25	**4)**	186	**5)**	114	**6)**	45
7)	18	**8)**	34	**9)**	32	**10)**	13	**11)**	21	**12)**	30

Exercises 42

1)	61	**2)**	13	**3)**	364	**4)**	387	**5)**	9	**6)**	36

Exercises 43

1)	85r1	**2)**	245r1	**3)**	158	**4)**	63r3	**5)**	92	**6)**	181r4

Exercises 44

1)	104r2	**2)**	62	**3)**	58r3	**4)**	128	**5)**	60r6	**6)**	136r5
7)	402	**8)**	250	**9)**	74	**10)**	312r2	**11)**	86	**12)**	19r37

Exercises 45

1)	23	**2)**	2	**3)**	365	**4)**	615	**5)**	7	**6)**	433
7)	7	**8)**	368								

Exercises 46

1)	8	**2)**	1,648	**3)**	623	**4)**	124r5	**5)**	347r6	**6)**	1264
7)	749	**8)**	157	**9)**	743						

Exercises 47

1)	1,200	**2)**	50 or 60	**3)**	80 or 90

Exercises 48

1)	26	**2)**	126	**3)**	388,716	**4)**	3	**5)**	1,071

CHAPTER 2

Exercises 1

1) One hundredth
   Expanded Form: $\frac{1}{100}$

2) Six tenths
   Expanded Form: $\frac{6}{10}$

3) Ten and eight hundredths
   $10 + \frac{8}{100}$

4) Nineteen and sixty-four hundredths
   $10 + 9 + \frac{6}{10} + \frac{4}{100}$

5) Five tenths
   Expanded Form: $\frac{5}{10}$

6) Four and twelve hundredths
   $4 + \frac{1}{10} + \frac{2}{100}$

7) Four and twelve thousandths
   $4 + \frac{1}{100} + \frac{2}{1000}$

Exercises 2

1)	9.05	2)	0.6	3)	0.26	4)	0.1	5)	7.34	6)	61.2
7)	0.81	8)	1.11								

Exercises 3

1)	>	2)	<	3)	<	4)	>	5)	>	6)	>
7)	<	8)	<	9)	>	10)	>	11)	>	12)	>

Exercises 4

1) $2.27 > 2.17 > 2.15 > 2.07 > 2.05$

2) $0.392 > 0.387 > 0.383 > 0.38 > 0.359$

3) $2.939 > 2.6489 > 2.648 > 1.998$

4) $0.65 > 0.6 > 0.05 > 0.35 > 0.3$

5) $0.26 > \frac{1}{4} > 0.245 > \frac{1}{5}$

6) $0.34 > \frac{1}{3} > 0.303 > 0.3 > \frac{3}{100}$

7) $10.01 > 9.01 > 8.90 > 0.99$

8) $2.85 > 2.65 > 2.58 > 2.56 > 2.36$

9) $0.10 > 0.05 > -0.03 > -0.05$

10) $0.22 > 0.11 > 0.05 > -0.11 > -0.22$

Exercises 5

1)	4.1	2)	4.4	3)	4.7	4)	4.9	5)	19.2	6)	19.6
7)	20.6	8)	20.8	9)	-2.5	10)	0.2	11)	0.9	12)	2.3

13) - 16)

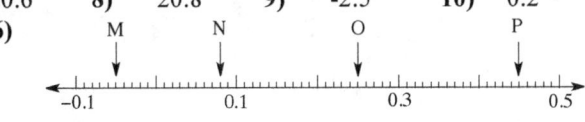

Exercises 6

1)	1.8	2)	1.1	3)	1.30	4)	2.1	5)	1.1	6)	6.30
7)	4.0	8)	9.1	9)	1.31	10)	2.25	11)	2.0	12)	1.73
13)	8.12	14)	16.05								

Exercises 7

1)	0.5	2)	0.13	3)	2.2	4)	0.56	5)	0.53	6)	1.3
7)	1.03	8)	3.75								

Exercises 8

1)	1	2)	10.2	3)	9.25	4)	1.01	5)	0.85	6)	1.80

Exercises 9

1)	10.86	2)	2.22	3)	0.39	4)	15.89	5)	22.3	6)	10.15
7)	10.24	8)	1.23	9)	8.32						

Exercises 10
1) 2.3   2) 1.18   3) 0.32   4) 3.56   5) 3.35   6) 0.23
7) 1.6   8) 1.27   9) 0.68   10) 1.84

Exercises 11
1) 0.75   2) 2.85   3) 0.4   4) 0.3   5) 0.5   6) 0.5
7) 0.46   8) 0.17   9) 3.58   10) 1.9   11) 0.75   12) 0.29
13) 8.52   14) 9.56   15) 3.45   16) 0.26   17) 0.88

Exercises 12
1) 0.5   2) 1.3   3) 1.6   4) 3.02   5) 2.87   6) 10.3

Exercises 13
1) 0.44   2) 1.82   3) 1.61   4) 0.29   5) 0.16   6) 0.92

Exercises 14
1) \$0.18   2) \$0.40   3) 2.08   4) 6.99   5) 0.44   6) 0.5
7) \$9.31   8) \$2.04   9) 38.29   10) 14.21   11) 2.33   12) 1.45

Exercises 15
1) 0.36   2) 2.38   3) 5.37   4) 2.93   5) 0.56   6) 9.78
7) 4.87   8) 8

Exercises 16
1) \$0.46   2) \$1.40   3) 5.95   4) \$0.20   5) 8.77   6) \$3.85
7) 2.22   8) 5.45   9) 0.25   10) 11.12   11) 0.38

Exercises 17
1) 0.14   2) 1.31   3) 6.55   4) 4.94   5) 2.91   6) 1.08
7) 11.41   8) 2.4

Exercises 18
1) $2 \times 2 \times 2$ 2) 2×17 3) 5×1 4) $2 \times 2 \times 5$

Exercises 19
1) 2   2) 17   3) 71   4) 2, 3, 7

Exercises 20
1) $2 \times 2 \times 2 \times 3$ 2) 2×3^2 3) 5^3 4) 31
5) 3×3 6) $2 \times 2 \times 2 \times 29$

Exercises 21
1) 84   2) 33   3) 30   4) 48   5) 36   6) 30

Exercises 22
1) 6   2) 12   3) 30   4) 4   5) 56   6) 12

Exercises 23
1) 12   2) 10   3) 10   4) 15

Exercises 24
1) $1\frac{1}{3}$ 2) $(\frac{1}{2})$ 3) $(\frac{3}{10})$ 4) $1\frac{1}{2}$ 5) $(\frac{1}{6})$ 6) $(\frac{1}{2})$

7) $2\frac{3}{8}$ **8)** $(\frac{4}{5})$

Exercises 25
1) One half   **2)** Two thirds   **3)** One eighth   **4)** One forth
5) One tenth   **6)** One hundredth

Exercises 26
1) $(\frac{1}{10})$ **2)** $(\frac{5}{10})$ **3)** $(\frac{4}{7})$ **4)** $(\frac{11}{13})$ **5)** $(\frac{8}{9})$ **6)** $(\frac{1}{100})$
7) $1\frac{1}{4}$ **8)** $2\frac{4}{5}$ **9)** $(\frac{3}{5})$ **10)** $1\frac{1}{2}$ **11)** $(\frac{1}{25})$ **12)** $(\frac{7}{8})$

Exercises 27
1) $(\frac{2}{6})$ **2)** $(\frac{2}{4})$ **3)** $(\frac{18}{24})$ **4)** $(\frac{6}{15})$ **5)** $(\frac{16}{28})$ **6)** $(\frac{12}{8})$

Exercises 28
1) $\frac{6}{14}, \frac{12}{28}$ **2)** $\frac{8}{6}), \frac{16}{12}$ **3)** 5, 45   **4)** 3, 45   **5)** 6   **6)** 30
7) 12   **8)** 5   **9)** 10   **10)** 3

Exercises 29
1) >   **2)** >   **3)** <   **4)** =   **5)** <   **6)** =
7) <   **8)** =   **9)** <   **10)** <   **11)** =   **12)** >
13) <   **14)** =

Exercises 30
1) $\frac{1}{2} > \frac{1}{3} > \frac{1}{4} > \frac{1}{5}$ **2)** $\frac{2}{1} > \frac{3}{2} > \frac{4}{3} > \frac{5}{4}$
3) $\frac{2}{2} > \frac{3}{4} > \frac{4}{6} > \frac{5}{8}$ **4)** $\frac{3}{4} > \frac{1}{2} > \frac{2}{6} > \frac{2}{8}$
5) $\frac{3}{4} > \frac{2}{3} > \frac{1}{2} > \frac{1}{8}$ **6)** $\frac{3}{4} > \frac{1}{2} > \frac{3}{5} > \frac{1}{3}$
7) $\frac{2}{5} > \frac{1}{3} > 0.30 > 0.25$ **8)** $0.75 > \frac{2}{3} > \frac{5}{8} > \frac{4}{7}$
9) $\frac{4}{6} > \frac{2}{4} > \frac{3}{9} > \frac{5}{15}$ **10)** $\frac{4}{5} > \frac{3}{4} > \frac{2}{3} > \frac{1}{2}$

Exercises 31

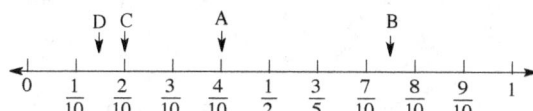

Exercises 32
1) $(\frac{2}{10})$ **2)** $(\frac{4}{10})$ **3)** $(\frac{8}{10})$ **4)** $(\frac{9}{10})$

Exercises 33
1) $(3\frac{1}{2})$ **2)** $(6\frac{2}{10})$ **3)** $(6\frac{9}{10})$ **4)** $(8\frac{3}{10})$

Exercises 34
1) $(7\frac{1}{3})$ **2)** 8   **3)** $(8\frac{1}{3})$ **4)** 9   **5)** $10\frac{1}{10}$ **6)** $10\frac{3}{10}$
7) $10\frac{6}{10}$ **8)** $10\frac{9}{10}$

Exercises 35

1) $(1\frac{1}{9})$ 2) 1 3) $(\frac{2}{3})$ 4) $(\frac{3}{5})$ 5) $(\frac{1}{2})$ 6) $(1\frac{2}{5})$

7) $(2\frac{3}{5})$ 8) $(2\frac{1}{2})$ 9) $(\frac{7}{8})$ 10) $(\frac{1}{3})$ 11) $(\frac{3}{4})$ 12) $(\frac{5}{8})$

Exercises 36

1) 1 2) $(\frac{3}{5})$ 3) $(\frac{1}{2})$ 4) $(\frac{2}{3})$ 5) $(1\frac{1}{3})$ 6) $(\frac{1}{8})$

Exercises 37

1) $(1\frac{2}{5})$ 2) $(3\frac{1}{3})$ 3) $(1\frac{1}{2})$ 4) $(5\frac{1}{2})$ 5) 1 6) 4

7) $(1\frac{3}{4})$ 8) $(5\frac{1}{3})$ 9) $(2\frac{3}{10})$ 10) $(5\frac{1}{5})$

Exercises 38

1) $(\frac{1}{2})$ 2) $(\frac{3}{4})$ 3) $(1\frac{4}{5})$ 4) $(1\frac{1}{2})$ 5) $(\frac{1}{9})$ 6) $(\frac{1}{2})$

7) $(\frac{4}{5})$ 8) $(1\frac{1}{2})$

Exercises 39

1) $(\frac{1}{2})$ 2) $(\frac{1}{2})$ 3) $(\frac{1}{4})$ 4) $(\frac{3}{4})$ 5) $(\frac{2}{9})$ 6) 0

7) $(\frac{1}{3})$ 8) $(\frac{7}{8})$ 9) $(1\frac{1}{7})$ 10) $(\frac{3}{5})$

Exercises 40

1) $(\frac{1}{4})$ 2) $(\frac{3}{8})$ 3) 1 4) 1 5) $(\frac{1}{2})$ 6) 2

7) $(1\frac{2}{9})$ 8) $(3\frac{2}{3})$

Exercises 41

1) $(2\frac{3}{4})$ 2) $(2\frac{7}{8})$ 3) $(1\frac{1}{3})$ 4) $(\frac{1}{3})$ 5) $(3\frac{1}{7})$ 6) 1

7) $(\frac{1}{4})$ 8) $(1\frac{1}{6})$ 9) $(\frac{1}{3})$ 10) $(\frac{1}{3})$ 11) $(\frac{2}{7})$ 12) $(\frac{1}{2})$

Exercises 42

1) $(\frac{1}{2})$ 2) 1 3) $(\frac{3}{5})$ 4) 1 5) $(\frac{1}{8})$ 6) $(\frac{1}{3})$

Exercises 43

1) 44.4 2) $1\frac{3}{8}$ 3) $(\frac{9}{16})$ 4) $(\frac{3}{8})$ 5) 11.1

Exercises 44

1) 0.5 2) 0.1 3) 0.375 4) 0.07 5) 0.86 6) 5.75

7) 0.67 8) 0.33 9) 0.08 10) 0.25 11) 0.75 12) 2.6

Exercises 45

1) $(\frac{1}{4})$ 2) $(\frac{7}{10})$ 3) $(\frac{14}{25})$ 4) $(5\frac{1}{5})$ 5) $(\frac{339}{100})$ 6) $(8\frac{101}{1000})$

7) $(\frac{3}{4})$ 8) $(\frac{7}{1000})$ 9) $(4\frac{9}{50})$ 10) $(1\frac{7}{25})$ 11) $(2\frac{3}{10})$ 12) $(2\frac{7}{50})$

Exercises 46

1) $(\frac{4}{5})$ 2) $(2\frac{1}{2})$ 3) 0.71 4) 0.04 5) $(\frac{3}{10})$ 6) 2.88

7) $(\frac{11}{50})$ 8) $(2\frac{1}{20})$ 9) 0.5 10) 0.101 11) $(10\frac{9}{20})$ 12) 2.07

CHAPTER 3

Exercises 1
1) 10 2) 3 3) 18 4) 8 5) 4

Exercises 2
1) $4 \div 2 - 2, 0$ 2) $(3 + 6) \times 2 + 3, 22$ 3) $(4 - 1) \times 5, 15$ 4) $(5 + 3) + (15 \div 5), 11$

Exercises 3
1) 107 2) 42 3) 19 4) 50 5) 12 6) 8
7) 0 8) 2 9) 8 10) 10 11) 112 12) 112

Exercises 4
1) (+) 2) (×) 3) 4 4) 1 5) 3 6) 9
7) 2 8) 27

Exercises 5
1) 2 2) 25 3) 5 4) 7.5 5) 4 6) 13
7) 10 8) 4 9) 0 10) 1 11) 15 12) 2

Exercises 6
1) 6.5 2) 2 3) 3 4) 16 5) 10 6) 2
7) 2 8) 5 9) (+) 10) 4

Exercises 7
1) 17, 10 2) 193 3) 18 4) 12 5) 11 6) 48
7) 12

Exercises 8
1) 12 2) 27 3) 42 4) 20 5) 8 6) 14
7) 17 8) 47 9) 60 10) 26 11) 44 12) 68

Exercises 9
1) 35 2) 208 3) 52 4) 61 5) 27 6) 59
7) 38 8) 81 9) 118 10) 22 11) 25 12) 14

Exercises 10
1) 15 2) 2 3) 18 4) 13 5) 17 6) 6

Exercises 11
1) 54 2) 77 3) 52 4) 45 5) 50 6) 26
7) 51 8) 61 9) 21 10) 229 11) 145 12) 38

Exercises 12
1) 9 2) 11 3) 12 4) 6 5) 3 6) 14

Exercises 13
1) 8 2) 34 3) 2 4) 4 5) 78 6) 29
7) 76 8) 67 9) 39 10) 14 11) 10 12) 4

Exercises 14
1) 30   2) 16   3) 8   4) 39   5) 22   6) 7
7) 29   8) 19   9) 62   10) 38   11) 95   12) 54

Exercises 15
1) 2   2) 4   3) 2   4) 27   5) 4   6) 15

Exercises 16
1) 32   2) 37   3) 42   4) 7   5) 87   6) 46

Exercises 17
1) 1   2) 2   3) 4   4) 22   5) 10   6) 15

Exercises 18
1) 12   2) 3   3) 8   4) 37   5) 19   6) 29

Exercises 19
1) 7   2) 12   3) 7   4) 11   5) 8   6) 8
7) 12   8) 8   9) 7   10) 5   11) 9   12) 7

Exercises 20
1) 72   2) 62   3) 7   4) 26   5) 13   6) 3

Exercises 21
1) 3.5   2) 7   3) 8   4) 4   5) 16   6) 2
7) 8   8) 13   9) 8   10) 4   11) 4

Exercises 22
1) 36   2) 4   3) 5   4) 18   5) 6   6) 16
7) 16   8) 2

Exercises 23
1) 6   2) 7   3) 18   4) 2   5) 7   6) 24

Exercises 24
1) 45   2) 7   3) 3   4) 12   5) 5   6) 63
7) 8   8) 15   9) 13   10) 6   11) 14   12) 7

Exercises 25
1) 1   2) 6   3) 22   4) 6   5) 13   6) 10
7) 4   8) 2

Exercises 26
1) 36   2) 64   3) 7   4) 7   5) 35   6) 32
7) 12   8) 15   9) 9   10) 10   11) 6   12) 15

Exercises 27
1) 2   2) 3   3) 7   4) 2   5) 4   6) 13
7) 12   8) 10

Exercises 28
1) 27   2) 45   3) 16   4) 160   5) 68   6) 21
7) 6   8) 24   9) 464   10) 6   11) 65

Exercises 29
1)   1   **2)**   3   **3)**   4   **4)**   5   **5)**   6   **6)**   4
7)   0.5   **8)**   3

Exercises 30
    $41.40

CHAPTER 4

Exercises 1
1)   B   **2)**   D   **3)**   F   **4)**   C   **5)**   A   **6)**   E

Exercises 2
1)   28 ft   **2)**   38 m   **3)**   24 cm   **4)**   72 in.   **5)**   25 ft   **6)**   44 cm
7)   70 in.   **8)**   28 cm

Exercises 3
1)   C   **2)**   A   **3)**   D   **4)**   B

Exercises 4
1) 3456 cm^3 **2)** 343 in^3 **3)** 720 mm^3 **4)**   5 cm
5)   11 ft   **6)** 72 ft^3 **7)**   4 yards   **8)**   12 in
9) 224 ft^3 **10)**

Exercises 5
1) 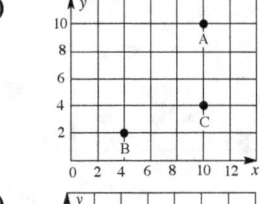   2) 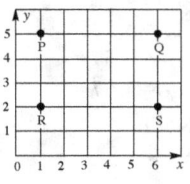   3)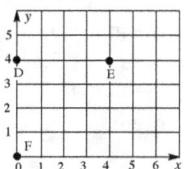

7) 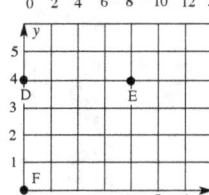   8) 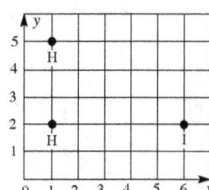   9)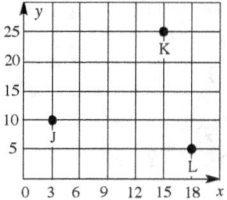

Exercises 6
1)   (6, 9)   **2)**   (15, 12)   **3)**   (0, 10)   **4)**   (6, 25)   **5)**   (18, 5)

Exercises 7
1)   5 units   **2)**   15 units   **3)**   6 units   **4)**   15 units

CHAPTER 5

Exercises 1

1)

Favorite Pet	Tally	Frequency
Fish	IIII	4
Bird	III	3
Dog	Ⱨ Ⱨ III	13
Cat	Ⱨ III	8
Hamster	I	1

2) dog   **3)** 5   **4)** 29

Exercises 2

1) 6   **2)** Vegetables   **3)** Hot dogs

Exercises 3

1) Dec.   **2)** About 1 h 15 m   **3)** Sep.   **4)** Oct.   **5)** The answer can vary.

Exercises 4

1) Mode = 8
Median = 11.5

2) Mode = 18
Median = 20.5

3) Mode ⁻ 20
Median = 20

4) Mode = no
Median = 66

5) Mode = no
Median = 11.5

6) Mode = 20
Median = 20

Exercises 5

1) Mean = 0.4625
Median = 0.475

2) Mean = 6.05
Median = 6.2

3) Mean = 32
Median = 30.5

4) Mean = 9.375
Median = 9.5

5) Mean = 5.125
Median = 5

6) Mean = 18.75
Median = 17.5

7) Mean = 137.5
Median = 158

8) Mean = 3.67
Median = 4

Self-Test, Page 10

1) B   **2)** C   **3)** A   **4)** C   **5)** A   **6)** A
7) C   **8)** A   **9)** D   **10)** A   **11)** C   **12)** D
13) D   **14)** C   **15)** D   **16)** C   **17)** A   **18)** D
19) B   **20)** B   **21)** C   **22)** B   **23)** D   **24)** D
25) A

Self-Test, Page 25

1) A   **2)** A   **3)** C   **4)** C   **5)** D   **6)** B
7) D   **8)** A   **9)** D   **10)** A   **11)** B   **12)** C
13) C   **14)** D   **15)** B   **16)** D   **17)** C   **18)** B
19) A

Self-Test, Page 42

1) B   **2)** C   **3)** D   **4)** C   **5)** D   **6)** B
7) B   **8)** A   **9)** D   **10)** D   **11)** B   **12)** C
13) A   **14)** C   **15)** B   **16)** A   **17)** B   **18)** C
19) D   **20)** A   **21)** C   **22)** B   **23)** B   **24)** C
25) C

Self-Test, Page 51
1)	B	2)	C	3)	A	4)	C	5)	A	6)	D
7)	C	8)	C	9)	C	10)	B	11)	C	12)	D

Self-Test, Page 60
1)	B	2)	B	3)	A	4)	C	5)	B	6)	C
7)	B	8)	C	9)	D	10)	C	11)	B	12)	C
13)	D										

Self-Test, Page 72
1)	D	2)	B	3)	D	4)	C	5)	C	6)	A
7)	B	8)	B	9)	A	10)	A	11)	B	12)	D
13)	D	14)	B	15)	D	16)	C	17)	D	18)	D
19)	B	20)	D	21)	B	22)	D	23)	A	24)	C
25)	D	26)	B	27)	B	28)	C	29)	D	30)	C

Self-Test, Page 84
1)	A	2)	A	3)	B	4)	D	5)	A	6)	B
7)	D	8)	B	9)	A	10)	C	11)	B	12)	C
13)	D	14)	B	15)	A	16)	B	17)	B		

Self-Test, Page 89
1)	D	2)	B	3)	B	4)	A	5)	D	6)	A
7)	D	8)	B	9)	B	10)	B	11)	A	12)	A
13)	B										

Self-Test, Page 95
1)	B	2)	C	3)	B	4)	D	5)	D	6)	C
7)	C	8)	D	9)	C	10)	D	11)	C	12)	B
13)	B	14)	A	15)	C	16)	C	17)	B	18)	C

Self-Test, Page 106
1)	C	2)	A	3)	D	4)	C	5)	B	6)	D
7)	A	8)	C	9)	B	10)	C	11)	A	12)	D
13)	B	14)	B	15)	C	16)	B	17)	A	18)	C

Self-Test, Page 116
1)	D	2)	C	3)	D	4)	C	5)	C	6)	D
7)	C	8)	D	9)	A	10)	B	11)	B	12)	B
13)	A	14)	B	15)	C	16)	D	17)	C	18)	A
19)	C	20)	B	21)	C	22)	A	23)	D	24)	B
25)	B										

Self-Test, Page 121
1)	B	2)	D	3)	C	4)	A	5)	D	6)	A
7)	D	8)	B								

Self-Test, Page 124
1)	D	2)	D	3)	A	4)	B	5)	C	6)	C
7)	C	8)	C	9)	C	10)	D	11)	D	12)	C

Self-Test, Page 126
1)	C	2)	D	3)	C	4)	B	5)	B	6)	C
7)	C	8)	C	9)	D	10)	B	11)	A	12)	C

13) B   14) A   15) B   16) A   17) C

Self-Test, Page 130
1) B   2) D   3) A   4) D   5) C   6) C
7) B   8) C   9) D   10) A   11) B   12) B
13) D   14) D   15) B   16) D   17) C   18) D

Self-Test, Page 136
1) D   2) D   3) A   4) B   5) C   6) C
7) C   8) B   9) C   10) D   11) A   12) C
13) D   14) D

Self-Test, Page 141
1) C   2) A   3) C   4) D   5) D   6) B
7) D   8) A   9) C   10) C   11) D   12) C
13) B

Self-Test, Page 149
1) C   2) C   3) B   4) A   5) D   6) C
7) C   8) D   9) D   10) A   11) C   12) B
13) A   14) D   15) C   16) A   17) D

Visit us at WWW.IQMATHS.com

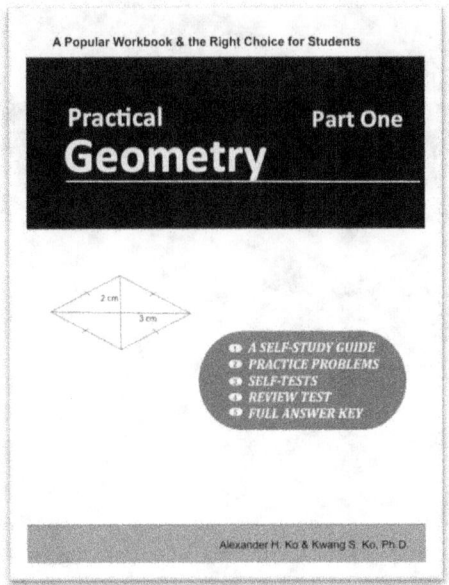

ISBN: 978-1-5232673-6-1

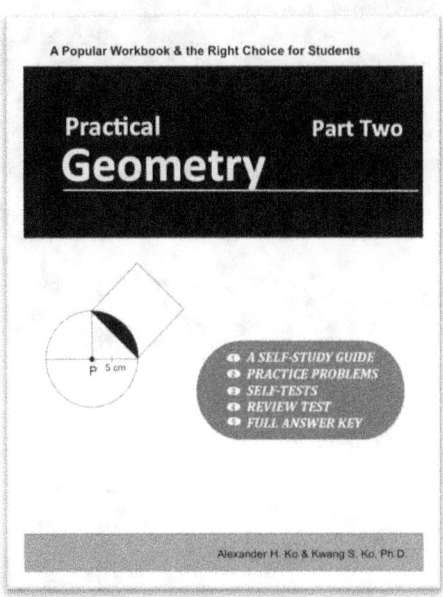

ISBN: 978-1-5233620-1-1

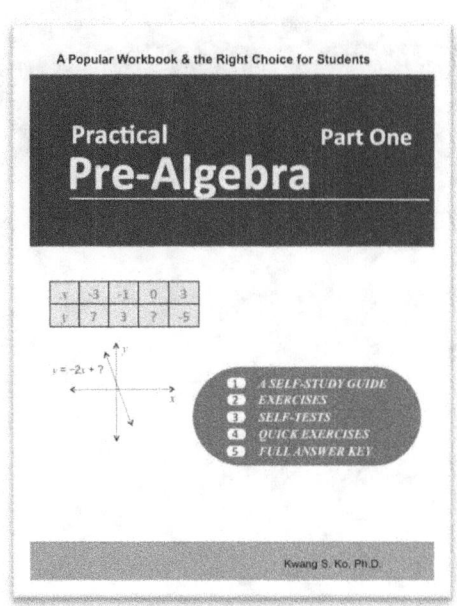

ISBN: 978-1-5233628-6-8

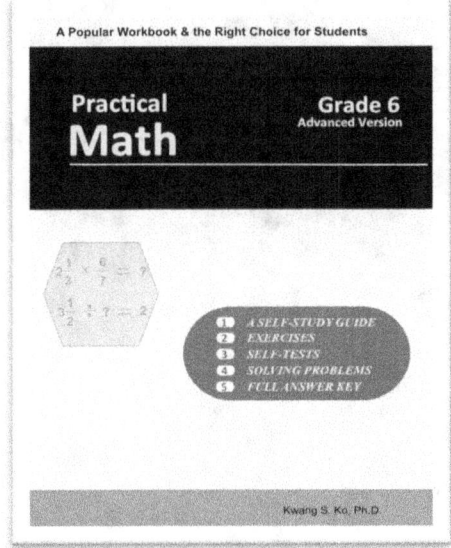

ISBN: 978-1-5233628-9-9

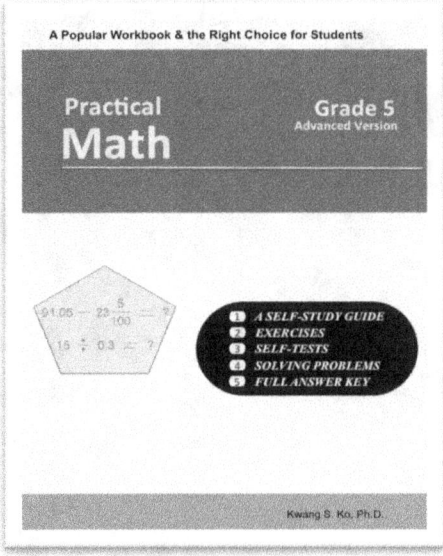

ISBN: 978-1-5233630-1-8

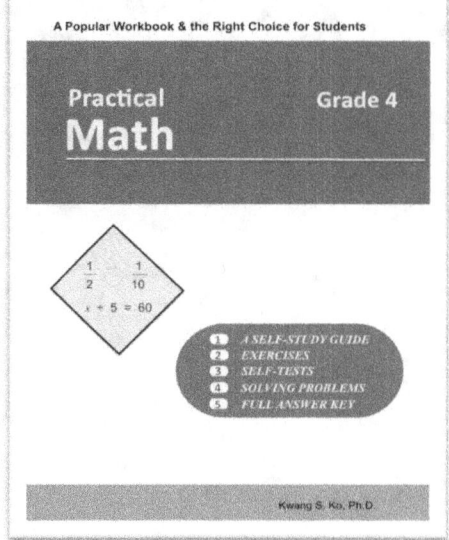

ISBN: 978-1-5233630-2-5

Other books are sold at WWW.IQMATHS.com.